PREFACE

Damage to earthen dams and dam safety issues associated with tree and woody vegetation penetrations of earthen dams is all too often believed to be a routine maintenance situation by many dam owners, dam safety regulators, and engineers. Contrary to this belief, tree and woody vegetation penetrations of earthen dams and their appurtenances have been demonstrated to be causes of serious structural deterioration and distress that can result in failure of earthen dams. For the first time in the history of dam safety, a Research Needs Workshop on Plant and Animal Impacts on Earthen Dams (Workshop) was convened through the joint efforts of the Federal Emergency Management Agency (FEMA) and the Association of State Dam Safety Officials (ASDSO) in November 1999 to bring together technical resources of dam owners, engineers, state and federal regulators, wildlife managers, foresters, and members of academia with expertise in these areas. The Workshop highlighted the realization that damage to earthen dams resulting from plant and animal penetrations was indeed a significant dam safety issue in the United States. The purpose of this *Technical Manual for Dam Owners, Impacts of Plants on Earthen Dams* is to convey technology assembled through the Workshop by successful completion of four objectives. These objectives are as follows:

1. Advance awareness of the characteristics and seriousness of dam safety problems associated with tree and woody vegetation growth impacts on earthen dams;

2. Provide a higher level of understanding of dam safety issues associated with tree and woody vegetation growth impacts on earthen dams by reviewing current damage control policies;

3. Provide state-of-practice guidance for remediation design considerations associated with damages associated with tree and woody vegetation growth on earthen dams; and

4. Provide rationale and state-of-practice techniques and procedures for management of desirable and undesirable vegetation on earthen dams.

ACKNOWLEDGEMENTS

The editors of this *Technical Manual for Dam Owners, Impacts of Plants on Earthen Dams* wish to acknowledge the support of the dam safety organizations and agencies and many dedicated individuals that made significant contributions to the contents of this Manual.

Sincere appreciation is extended to past and present members of the Subcommittee on Dam Safety Research of the Interagency Committee on Dam Safety (ICODS), now the National Dam Safety Review Board Work Group on Dam Safety Research, for their support of the proposal to convene a Research Needs Workshop on Plant and Animal Impacts on Earthen Dams (Workshop), and to the members of ICODS for recommending funding for the Workshop through the Federal Emergency Management Agency (FEMA).

Appreciation and sincere gratitude are extended to the members, and especially to the full-time staff, of the Association of State Dam Safety Officials (ASDSO) for their support and coordination of the federally-funded project that culminated with convening of the Research Needs Workshop on Plant and Animal Impacts on Earthen Dams at the University of Tennessee in Knoxville, Tennessee, on November 30-December 2, 1999. The editors of this Manual are especially appreciative of the continued support, patience, and dedication of Susan Sorrell and Sarah Mayfield of the ASDSO staff.

The Steering Committee of the Workshop was comprised of the following individuals who contributed significantly and diligently to making the Workshop a truly historical dam safety technological event:

Dr. B. Dan Marks, P.E. (Chairman)	**Charles Clevenger, MS (Deceased)**
Dr. Bruce A. Tschantz, UT Knoxville	**William L. Bouley, USBR**
David K. Woodward, NCSU	**Sarah M. Mayfield, ASDSO**

Susan A. Sorrell, ASDSO (Project Coordinator)

Participants in the Workshop brought together diverse technologies, experiences, and scientific developments to create a significant contribution to dam safety in the United States. The editors of this Manual acknowledge the valuable contributions of the following Workshop participants:

Matthew A. Barner, Wright State Univ.

William L. Bouley, USBR

Charles Clevenger, MS (Deceased)

Dr. Kim D. Coder, Univ. of Georgia

Gary Drake, Reemay, Inc.

Edward Fiegle, GA Dam Safety

James K. Leumas, NC Dam Safety

Dr. B. Dan Marks, Marks Enterprises

Sarah M. Mayfield, ASDSO

Dr. Marty McCann, NPDP-Stanford

Douglas E. McClelland, USDA Forest Service

Dr. James E. Miller, USDA-CSREES/NRE

Dr. Dale L. Nolte, USDA-APHIS/WS/NWRC

Richard D. Owens, USDA-APHIS/WS

Tom Renckly, Maricopa County, AZ

Dr. David Sisneros, USBR

Boris Slogar, OH Dam Safety

Susan A. Sorrell, ASDSO

Dr. Bruce A Tschantz, UT Knoxville

David K. Woodward, NCSU

The contents of this Manual were significantly enhanced by editorial reviews and comments of the ASDSO Manual Overview Committee. The editors of the Manual are sincerely appreciative of contributions made by the following individuals:

Timothy G. Schaal, SD Water Rights Program

R. David Clark, MA Office of Dam Safety

Daniel M. Hill, Burgess & Niple, Ltd.

James K. Leumas, City of Raleigh, NC

Lori Spragens, ASDSO Executive Director (Project Coordinator)

Because of the efforts of the many individuals previously mentioned, the editors are confident that users of this Manual will develop a better understanding and gain a greater appreciation of the seriousness and magnitude of problems associated with the effects of tree and woody vegetation root penetrations on the safety of earthen dams and their appurtenances.

TABLE OF CONTENTS

GLOSSARY

This glossary provides the definitions of some of the basic terms used in this *Manual* and is not intended to be a comprehensive glossary of terms associated with dam safety. A more extensive resource of dam safety terms and definitions is available through the many references provided at the end of each chapter of the *Manual*.

Absorption - the process of being taken into a mass or body, as water being taken in by plant roots.

Abutments - the interface between the sides of a valley containing a dam and the dam embankment. Right and left abutments are referenced by viewing the dam while facing downstream.

Adsorption – the adhesion of an extremely thin layer of molecules to the surface of solid bodies or liquids with which they are in contact.

Appurtenances – structures associated with dams such as spillways, gates, outlet works, ramps, docks, etc. that are built to allow proper operation of dams.

Berm – a horizontal step or bench in the embankment slope of an earthen dam.

Biological Barrier – an herbicidal releasing system, device, or material designed to exclude root growth and/or penetration of plants into a protected underground zone (such as a dam embankment).

Boil – a typically circular feature created by the upward movement of soil particles by seepage flowing under a pressure slightly greater than the submerged unit weight of the soil through which seepage is occurring.

Breach – a break, gap, or opening in a dam that typically allows uncontrolled release of impounded water.

Capillary Rise – the rise of water in the voids of a soil mass as a result of the surface tension forces of water.

Clearing – the removal of trees and woody vegetation by cutting without removal of stumps, rootballs, and root systems.

Crest – the near horizontal top surface of an earthen dam, or the control elevation of a spillway system.

Diameter at Breast Height (dbh) – the diameter of a tree measured at about four feet (breast height of average person) above the ground surface.

Drainage System – graded and/or protected pervious aggregates in a dam designed to collect, filter, and discharge seepage through the embankment, abutments, or foundation.

Earthen Dam – a dam constructed of compacted natural soil fill materials selected to minimize embankment seepage while maximizing workability and performance.

Embankment – an earthen or rockfilled structure having sloping sides constructed of select compacted fill materials.

Failure – a (dam) incident that results in the uncontrolled release of water from the impoundment of a dam.

Freeboard – the vertical distance from the normal operating water level of an impoundment to the crest (top) of the dam.

Grubbing – the removal of stumps, rootballs, and lateral root system of trees and woody vegetation. A construction operation that is typically done following the clearing operation.

Herbicide – a chemical substance or mixture designed to kill or maintain undesirable Plants that may include herbaceous plants, vines, brush, and trees.

Hydraulic Height (of a Dam) – the vertical distance from the normal operating water level of the impoundment to the invert of the outlet works or downstream outlet channel.

Hydro-seeding – the technique of applying grass seeds, fertilizer, agricultural lime, and seedbed mulch to seeded area in a pressurized aqueous mixture.

Lateral Root System – roots of trees and woody plants that extend laterally from the tap root and/or rootball to provide lateral support and nutrient uptake for the plant.

Line of Saturation – the leading boundary of the progression of saturation of soil in an embankment exposed to an increasing head (source) of water (impoundment).

Line of Wetting – the leading boundary of the progression of wetting (partial saturation) of soil in an embankment exposed to an increasing head (source) of water (impoundment).

Maintenance – routine upkeep necessary for efficient inspection, and safe operation and performance of dam and their appurtenances. Labor and materials are required; however, maintenance should never be considered to comprise dam remediation.

Mowing – the cutting of grass, weeds, and small-diameter woody vegetation by mechanical devices such as mowers, bush hogs, and other vegetation cutting machinery.

Mulching – the application of protective material such as straw, fiber matting, and shredded paper to newly seeded areas.

Operation (of a dam) – activity by a dam owner for the necessary and safe use and performance of a dam, the appurtenances of a dam, and the impoundment.

Owner – any person or organization that owns, leases, controls, operates, maintains, or manages a dam and/or impoundment.

Phreatic Surface – the upper boundary (surface) of seepage (water flow) zone in an embankment.

Piping – the progressive downstream to upstream development of internal erosion of soil as a result of excessive seepage that is typically observed downstream as a hole, or boil, that discharges water containing soil particles.

Remediation – restoration of a dam to a safe and intended design condition.

Revegetation – restoration of desirable ground cover vegetation (i.e. grasses) to disturbed areas designed to prevent embankment surface erosion.

Rootball – the root and soil mass portion of a tree or woody plant that is located directly beneath the trunk or body of the tree that provides the primary vertical support for the tree or woody plant.

Root Penetration – intrusion of plant roots into a dam embankment so as to interfere with the safe hydraulic or structural operation of the dam.

Root System – roots contained in the rootball and the lateral root system collectively comprise the root system of trees and woody plans and provide both lateral and vertical support for the plant as well as providing water and nutrient uptake for the plant.

Seeding – application of a seeding mixture to a prepared seedbed or disturbed area.

Seepage – the flow of water from an impoundment through the embankment, abutments, or foundation of a dam.

Seepage Line – the uppermost boundary of a flow net, or the upper surface (boundary) of water flow through an embankment (see Phreatic Surface).

Slump – a portion of soil mass on an earthen dam that has or is moved downslope, sometimes suddenly, often characterized by a head scarp and tension cracks on the crest and embankment slope.

Spillway Systems – control structures over or through which flows are discharged from the impoundment. Spillway systems include Primary or Principal Spillways through which normal flows and small storm water flows are discharged and Auxiliary or Emergency Spillways through which storm water flows (floods) are discharged.

Stripping – the removal of topsoil, organic laden materials, and shallow root systems by excavating the ground surface (surficial soil stratum) after grubbing an area.

Structural Height (of a Dam) – the vertical distance from the crest (top) of the dam to the lowest point at the toe of the downstream embankment slope, or downstream toe outlet channel.

Stump – that portion of the trunk or body of a tree or woody plant left after removal by cutting during timber harvesting and/or clearing of trees and woody plants.

Stump Diameter – the diameter of a tree or woody plant at the ground surface.

Tap Root – the primary vertical root in the rootball that is the origin of development for the rootball and lateral root system growth.

Toe of Embankment – the point of intersection of the embankment slope of a dam with the natural ground surface.

Weeds – shallow-rooted, non-woody plants that grow sufficiently high as to hinder dam safety inspections and do not provide desirable embankment slope protection against surface runoff.

Woody Vegetation – plants that develop woody trunks, rootballs, and root systems that are not as large as trees but cause undesirable root penetration in dams.

Zone of Aeration – the partially saturated zone of a soil mass above the zone of saturation (above the height of capillary rise of water in a soil mass).

Zone of Saturation – the saturated zone of a soil mass above the phreatic surface defined by the height of capillary rise.

Chapter 1
Introduction

At the time Joyce Kilmer dedicated his famous poem "**Trees**" to Mrs. Henry Mills Alden, he was undoubtedly inspired by the beauty of a healthy living tree, and rightly so. For those that do not remember, the first verse of this famous poem is as follows: "*I think that I shall never see / A poem lovely as a tree.*" Most people are inspired and impressed by the splendor of trees; however, dam owners, operators, inspectors, dam safety regulators, engineers, and consultants might find the following verse more nearly appropriate. "*I think that I shall never see / A sight so wonderful as a tree / Removed from an earthen dam / Whose future safety we wish to see.*" This paraphrased verse is not intended to debase the great works of Joyce Kilmer; but rather, is intended to draw attention to the fact that trees and woody vegetation growth have no place on the embankment of an earthen dam.

Dam safety regulators and inspectors, engineers, and consultants are frequently confronted with grass roots resistance in the issue of removal of trees and woody vegetation from earthen dams. This resistance is often associated with sentimental, cultural, ecological, legal, and financial issues. A fundamental understanding and technical knowledge of potential detrimental impacts of trees and woody vegetation growth on the safety of earthen dams is necessary in order to address these issues.

Purpose

The purpose of this *Manual* is to provide the dam owner, operator, inspector, dam safety regulator, engineer, and consultant with the fundamental understanding and technical knowledge associated with the potential detrimental impacts of tree and woody vegetation growth on the safety of earthen dams. In addition to objectives related to raising the knowledge level of detrimental effects of trees and woody vegetation growth on the safety of earthen dams, the contents will provide the user of this *Manual* with an understanding of the methods, procedures, and benefits of maintaining a growth of desirable ground covering vegetation on the embankments of earthen dams.

Scope

The editors of this *Manual* have organized the contents in a sequential manner in order that the reader and user of this *Manual* can develop the desired fundamental understanding and gain the technical knowledge associated with the detrimental impacts of tree and woody vegetation growth on earthen dams. Chapter 2 deals with the problems associated with tree and woody vegetation growth on earthen dams. Chapter 3 presents some common misconceptions about tree growth and tree root development. These misconceptions are contrasted with factual data about tree growth and tree root development.

Chapter 4 presents a recommended earthen dam inspection protocol and procedures for determination of potential impacts of tree and woody vegetation growth on earthen dams. Chapter 5 begins the presentation of proper vegetation management on earthen dams. The user of this *Manual* is presented with methods and procedures for maintaining desirable vegetation growth, while also controlling tree and woody vegetation growth.

Chapter 6 presents a number of remediation design considerations associated with the removal of trees and woody vegetation from the embankments of earthen dams. This chapter also presents a recommended phased-remediation procedure for removal of undesirable vegetation (trees and woody vegetation) from earthen dam embankments. Chapter 7 is a succinct factual presentation of costs associated with either continual proper vegetative maintenance or long-term dam remediation construction after tree and woody vegetation removal. The contents of this chapter should make every dam owner cognizant of the need for proper operation and maintenance relative to vegetative growth on earthen dams.

Implementation

While this *Manual* may not be considered highly technical relative to the presentation of complex engineering calculations for the solution of potentially serious earthen dam safety problems, this *Manual* does present a combined sixty-five years of research and practice in dam safety engineering associated with tree and woody vegetation growth impacts on earthen dams. This *Manual* is presented in a manner to be beneficial to the entire dam safety community (dam owners, dam operators, dam safety inspectors, dam safety regulators, dam safety engineers and consultants). Dam safety engineers and consultants can utilize this *Manual* as a reference for recommendations for   proper maintenance of desirable vegetation growth, control of undesirable vegetation growth, and remediation dam design associated with the removal and control of trees and woody vegetation growth on earthen dams. Dam safety regulators and dam safety inspectors can utilize this *Manual* as a guideline for the inspection of earthen dams relative to tree and vegetation growth dam safety issues and for the direction of dam owners and operators in the proper method and procedures for maintaining earthen dams without detrimental vegetative growth. Dam owners and operators can utilize this *Manual* to establish proper operation and maintenance programs to promote the growth of desirable vegetative growth on earthen dams and/or remove and control the undesirable tree and woody vegetation growth on earthen dams.

The last verse in the famous poem **Trees** by Joyce Kilmer is as follows: **"Poems are made by fools like me / But only God can make a tree."** Again, the author will paraphrase this last verse, not to debase the great works of Joyce Kilmer, but to make a distinct point. **"Only God can make a tree / But not removing trees from dams / Is done by fools like me."**

There is yet much research and study to be done relative to the growth of proper vegetative cover on earthen dams. However, there is no doubt that trees and woody vegetation have no place on the embankment slopes of an earthen dam. The authors of this *Manual* intend to continue technological development in the area of controlling tree and woody vegetation growth on earthen dams. The authors would appreciate documentation of unusual cases of tree and woody vegetation growth related to safety issues associated with earthen dams. Documentation of these issues can be communicated through ASDSO and/or directly to the authors of this *Manual*.

Chapter 2
Problems with Tree and Woody Vegetation Growth

According to the 1998-99 National Inventory of Dams (NID) data, there are approximately 76,700 dams of significant size[1] and hazard category in the 50 states (USCOE, 1999). Most of these dams are regulated by the jurisdictional states, but many are not because of specific exemption clauses or different size or hazard restrictions. Because some states have lower size definitions for their dams than used for the NID count, the actual number of state-regulated dams is much higher (about 94,000). In Tennessee over 40 percent of the approximately 1000 inventoried dams *not* subject to regulation because of statutorily named county exclusions or agricultural use exemptions. Most of these unregulated dams and some of the regulated dams in Tennessee have troublesome trees and brush growing on their faces and crests. Some states estimate that as many as 95 percent of their regulated dams have trees. Figure 1 illustrates the general magnitude and range of the tree growth on regulated dams in 48 states where this information is reported (ASDSO, 2000). About half of the state-regulated dams are estimated to have excessive tree growth.

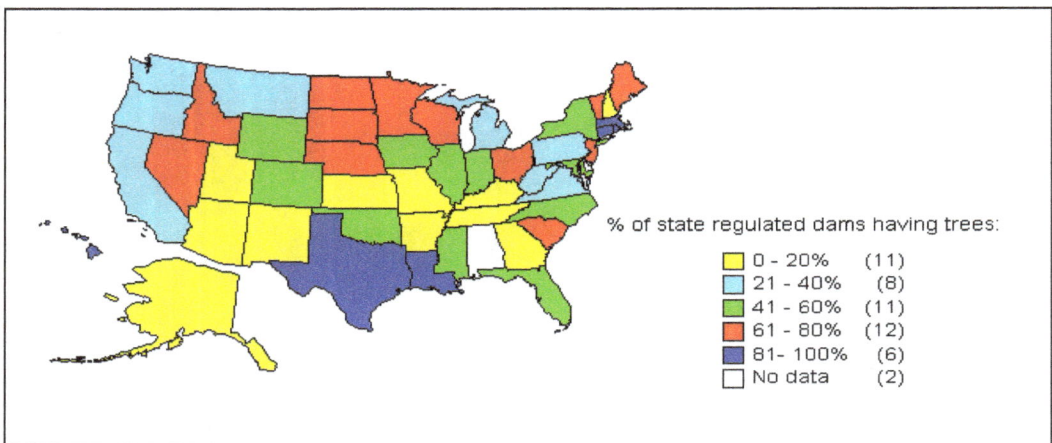

Figure 1. Estimated percentages of state-regulated dams having trees.

---

[1] Inclusion in the National Inventory has been defined under P.L. 99-662 and P.L. 92-367 to include dams that are at least 25 ft. high or 50 acre-feet of storage (excluding low hazard dams less than 6 ft. high or 15 acre feet of storage) and dams that due to location may pose a significant threat to human life or property in event of failure.

Most dam safety engineers, including state and federal officials, consultants, and other experts involved with dam safety, agree that when trees and woody plants are allowed to grow on earthen dams, they can hinder safety inspections, can interfere with safe operation, or can even cause dam failure. In the past, engineers and dam safety experts have not always been in agreement about the best way to prevent or control tree growth, remove trees, or repair safety-related damages caused by trees and woody vegetation. However, all dam engineers agree that a healthy, dense stand of low-growing grass on earthen dams is a desirable condition and should be encouraged.

From November 30 - December 2, 1999, a joint ASDSO/FEMA-sponsored workshop was held in Knoxville, Tennessee, for the purpose of inviting a panel of experts to discuss various problems, policies, and practices associated with plant and animal penetrations of earthen dams. Much of this manual follows up the work and recommendations produced by the workshop participants for engineers and owners to use in managing problems associated with both plant and animal intrusions. This chapter will discuss the consensus of current attitudes, issues, and policies involving woody vegetation penetrations of earthen dams, by state and federal officials, researchers, and practitioners active in dam safety.

Attitudes Toward Woody Plant Growth on Dams

The Association of State Dam Safety Officials (ASDSO) sent out survey questionnaires to dam safety officials in all 50 states and to federal representatives to the Interagency Committee on Dam Safety (ICODS) to determine state and federal agency attitudes about the effects of trees and woody plant growth on dams under their jurisdiction (ASDSO, 1999).

In this survey the state and federal agency representatives were asked (1) if they considered vegetative growth to be a problem on dams, (2) if they had specific policies or operating procedures for removing unwanted vegetation and trees on dams and if they didn't, how did they handle such problems, (3) what legal, financial, environmental or other constraints did they have in dealing with unwanted vegetation problems, (4) to provide documented evidence and examples where vegetation has negatively affected the safe operation or has contributed to the failure of dams, (5) to provide references to current or past research regarding the effects of plants and trees on dam safety, and (6) to provide example cost and other information related to rehabilitation and remediation of dams having problem woody plant growth. This chapter summarizes the collective state and federal attitude, and practice toward woody plant growth on dams.

Problems Caused by Trees and Woody Plants

Of the 48 states that responded to the above seven questions (Alabama and Delaware did not reply), all state dam safety officials indicated that they consider trees and plant growth on dams to be a safety problem. One eastern state dam safety engineer goes so far to say that trees are probably the major problem that he has to deal with. He notes further that most of the trouble occurs because owners (and some engineers) do not recognize trees as problems and become complacent as trees slowly grow into serious problems. Both state and federal officials agree that trees have no place on dams. Federal agencies like the Corps of Engineers, U. S. Bureau of Reclamation, and TVA, which own, operate and maintain their own dams, do not allow trees to grow on their structures. Figure 2 shows a problem dam in Nebraska where tree roots have been reported to penetrate the chimney drain and thus affect the operation of the dam.

Figure 2. Example dam with problematic trees in Nebraska.

Figure 3. Example dam with inspection-hindering trees in Tennessee.

The problem most commonly noted by state officials is that trees, woody vegetation, briars, and vines interfere with effective safety inspections. Figure 3 illustrates this problem for a dam located in Tennessee.

Figure 4 gives a breakdown of the percentage ranges of regulated dams where the 48 reporting state dam safety officials shown in Figure 1 estimate that trees and brush hinder safety inspections in their respective states (ASDSO, 1999). While half the states report having only 20 percent or fewer dams with significant trees and woody vegetation that hinder inspections, vegetation on an estimated 30,000 or nearly a third of the collective state-regulated dams, is reported to obstruct effective dam safety inspections.

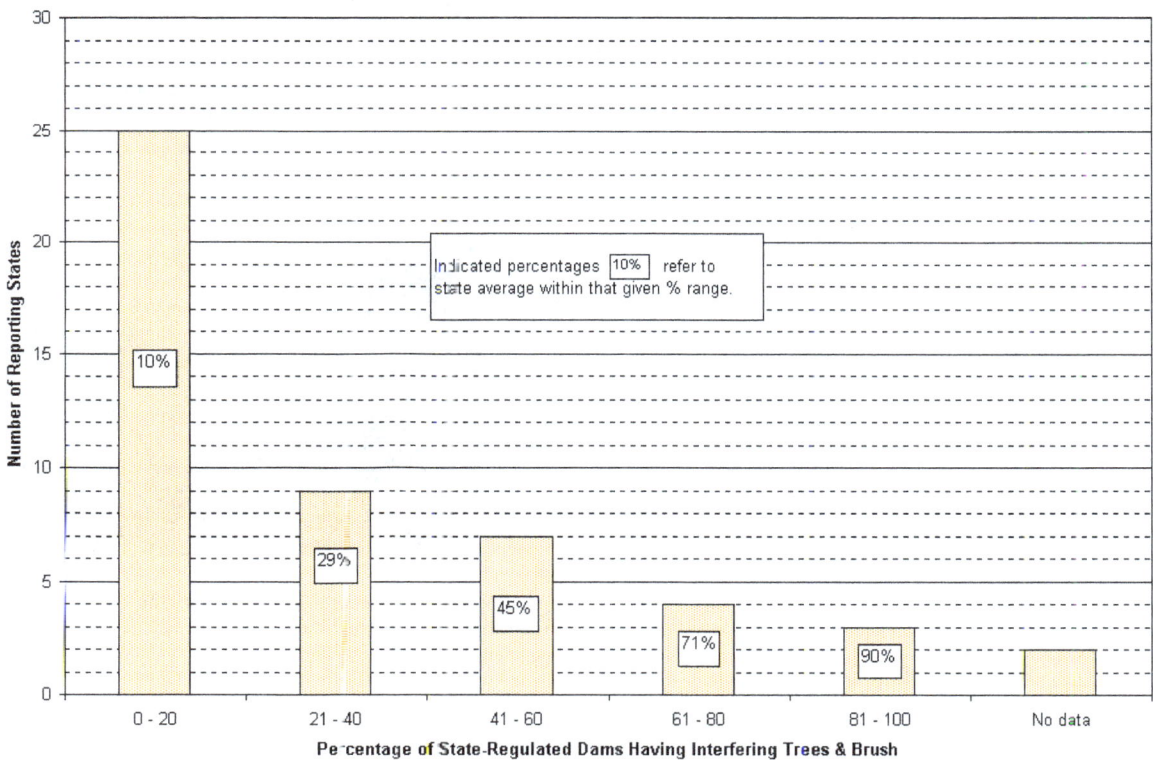

Figure 4. Estimated percentages of state-regulated dams where trees and brush are considered a deterrent to effective safety inspections.

Other dam safety problems caused by woody vegetation growth are:

- Uprooted trees that produce large voids and reduced freeboard; and/or reduced x-section for maintaining stability as shown in Figure 5.
- Decaying roots that create seepage paths and internal erosion problems.
- Interfering with effective dam safety monitoring,

Figure 5. Serious damage by uprooted tree to embankment stability at a dam in Oregon.

  inspection and maintenance for seepage, cracking, sinkholes, slumping, settlement, deflection, and other signs of stress

- Hindering desirable vegetative cover and causing embankment erosion
- Obstructing emergency spillway capacity
- Falling trees causing possible damage to spillways and outlet facilities
- Clogging embankment underdrain systems
- Cracking, uplifting or displacing concrete structures and other facilities

Figure 6. Tree root induced scouring on crest and downstream face of Coffey dam in Kansas.

- Inducing local turbulence and scouring around trees in emergency spillways and during overtopping as shown in Figure 6.
- Providing cover for burrowing animals
- Loosening compacted soil
- Allowing roots to wedge into open joints and cracks in foundation rock along abutment groins and toe of embankment, thus increasing piping and leakage potential.
- Root penetration of conduit joints and joints in concrete structures

Current Policies and Procedures

Twenty-four of the 48 responding states noted that they had formal policies and/or operating procedures for addressing tree and woody plant growth issues. These policies usually include one, or some combination, of the following:

- Trees are not allowed to grow on dams or near toe and abutment
- All trees and stumps must be removed, but roots may be left
- All trees, stumps, and roots must be removed
- All trees must be removed, but root systems of "small" trees may be left; root systems of "large" trees must be removed
- Dams are treated on a case-by-case basis -- usually under the direction of a qualified professional engineer.

For those states that choose to distinguish between "small" and "large" trees, the definition basis ranges from two to eight inches in diameter; most use a size of four or six inches in carrying out their policies.

Of the remaining 24 states indicating that they have no formal policies or procedures, the range of recommended procedures to dam owners varies widely. Some states evaluate dams on a case-by-case basis, while other states require owners either to maintain their dams, to remove vegetation for inspection, or to use other means for dealing with plant problems such as requiring a qualified engineer to be retained, depending on the dam hazard classification.

In summary, states follow several schools of thought and considerations in dealing with trees and vegetation on existing and new dams:

Existing Dams:

- Distinguish between "small" trees and "large" trees
- Remove all trees, stumps, and roots from dam embankment
- Cut trees to ground level, but leave stumps and roots
- Cut trees, remove stumps, but leave roots
- Consider case-by-case basis
- Breach, remove, or decommission dam
- Require retention of a qualified engineer by owner
- Do nothing.

Figure 6. Trees cut prior to removing stumps and roots from dam.

Chapter 4: Dam Remediation Design Considerations presents recommended procedures for removal of trees and dealing with tree and woody vegetation related problems.

Figures 6 and 7 illustrate extensive efforts necessary to restore a heavily wooded earthen dam to a desirable vegetated and maintained condition.

Figure 7. Completed remediation job after removing stumps, seeding, fertilizing & mulching.

New Dams:

- Establish effective ground cover and hope for the best in continual maintenance
- Use vegetative barriers such as bio-barriers, or use silvicides/herbicides/chemical treatment.

Constraints to Removing Trees and Plants

Several state and federal dam safety officials reported constraints to removing and/or controlling unwanted trees and other vegetation. Constraint categories explicitly cited by state dam safety officials (number of states in parentheses) are given below:

- Financial limitations by owners (13 states)
- Environmental regulations and/or permits (10 states)
- Legal issues (6 states)
- Aesthetics (5 states)
- Threatened/endangered species issues (2 states)
- Media (1 state)
- Sentimental reasons (several).

States indicated that the greatest constraint to removing unwanted trees and plants and repairing a structure infested with roots is limited financial capability by the owner. States such as Kentucky try to work with the owner to minimize the financial burden without threatening public safety. Ohio has recently established two low-cost loan programs to assist qualified public and private dam owners in funding safety-related improvements to their dams, including repairs mandated by the state dam safety program.

Environmental constraints range from limitation of the use of certain herbicides or chemicals for controlling vegetation and for treating stumps and/or roots near water bodies; to prohibition of, or air quality concern for, burning cleared vegetation. Unless exempted, vegetation removal and maintenance around dams may conflict with wetland protection regulations. In Washington, environmental issues can pose a major hurdle to removing trees, but ultimately, public safety takes precedence over environmental concerns. In Arizona, problems with time-consuming environmental permit requirements for larger plant removal projects are sometimes encountered.

Some states have limited legal power to force owners to remove trees and vegetation from dams. This lack of authority may cause delays and expensive and time-consuming litigation to obtain an order. Other states, like Maine, do not have specific laws that force owners to remove vegetation from their dams, and removal orders have yet to be tested. One state, South Carolina, notes that if the owner will not voluntarily cut or remove unwanted vegetation, the only course is to start legal action against the owner. Because legal help is limited, such help is normally requested for the "most extreme cases." This means that only a few owners can be forced to do something about their vegetation. In New Hampshire, legal assistance is sometimes necessary to perform enforcement functions. In Oregon, if there is a problem with a recalcitrant owner, a Proposed Order can be initiated by the Oregon Dam Safety Program to correct the situation if it is determined to be an immediate threat to the integrity of the structure. However, this process can be rather lengthy and expensive when staff time, materials, and attorney fees are included in the costs of preparing for a contested case hearing. In the end, most dam owners have the right to contest state directives to remove trees and other plants through administrative and legal processes and judicial appeals.

In some states, concerns have arisen when dams are located in parks or environmentally sensitive areas, especially when endangered or threatened species habitat is involved, in turn creating legal constraints.

Aesthetics and sentimental reasons are often used by dam owners and their neighbors to resist removing trees and undesirable vegetation. This is particularly true if owners have intentionally planted ornamental trees and shrubs on their dams to provide shade or fruit, or to improve looks. Some owners believe that the more woody vegetation on a structure, the better -- thus making it very difficult for state dam safety officials to request its removal.

The power of the press has had major influence on tree removal programs in some cases, especially where the target dam is owned by a poor or downtrodden citizen or insolvent municipality. Heated controversy between public safety interests and private owners or interest groups was generated through various newspaper stories and letters to the editor in 1990 over the removal of 500 mature cottonwood trees on two dams owned by an 85-year-old widowed rancher who at the time was suffering from serious illness. The news stories, which cast the owner as being targeted because she was vulnerable, influenced the owner's neighbors to encourage her to take a stand against further removal of 500 remaining trees because they felt that enforcement of the state dam safety act "would cause more harm than good."

While these constraints affect the ability of many states to enforce their regulations, some states such as Arkansas, Georgia, Colorado, Iowa, Maryland, Montana, New Jersey, North Carolina, and Tennessee report no major constraints to enforcement and consider the safety of dams to be of primary importance.

Federal agencies appear to have fewer constraints than states relative to mandating the upkeep and maintenance of jurisdictional dams. However, some federal agencies noted that they must make sure that they comply with the National Environmental Policy Act and the Endangered Species Act prior to initiating tree and plant control and management. Isolated constraints at the National Park Service involving funding priorities, historic preservation, and disruption of visitor services may override safe operation and maintenance needs at some dams. Local watershed districts that are often poorly funded are responsible for the operation and maintenance of many of the USDA/NRCS flood control dam projects.

Vegetation-Caused Problems and Failures

Twenty-nine states indicated documented evidence where vegetation on dams has either caused dam failure or negatively affected their safe operation. Sixteen states had no documented evidence and five states had no response. Several states provided photos (Figure 8) and information on tree caused failures or dam

Figure 8. Exposed tree roots in overtopped dam.

safety problems.   The most recent documented dam failure due to tree root penetration occurred in May 1999 at an unnamed Air Force Academy dam near Colorado Springs.   Here, an approximately 13-ft. high dam with a pond capacity of less than 5 acre-feet of horse stable waste water failed, releasing its contents and injuring a horse in a stable located about 100 yards downstream.   The failure occurred after more than 7 inches of rain had fallen in the previous 72 hours.   The dam had several pine trees on its crest and faces, and the breach opening exposed an extensive, deep root system.   Roots up to 4 inches in diameter were found in the breach area.   Figure 9 shows an example of a large root exposed in the bottom of the channel at the breach. The dam had not overtopped, and the failure was attributed to internal erosion of the decomposed granite embankment material along the roots.   A tree had been located directly over the breach.

Figure 9. Large pine tree root located in the channel of the breach opening of a failed
Air Force Academy waste lagoon pond dam (David Eyre, Senior Civil
Engineer, Air   Force Academy, Colorado, 1999).

At the Federal level, USDA/NRCS referred to documented cases where dam failure has been determined to be caused solely by trees, and noted that trees have also masked other more serious seepage problems, which went undetected.

Past and Current Research

Other than a few references to the University of Tennessee Tree Growth Report (Tschantz, 1988), only one or two other citations for tree or woody plant-related research were identified by the state dam safety officials (USDA/SCS, 1981). The surveyed Federal agencies had relatively little to offer in the way of references to current or past research regarding the effects of tree and plant growth on dam safety. The Corps of Engineers referred to geotechnical and other related program research conducted at the Waterways Experiment Station, published as a technical report series, Repair- Evaluation-Maintenance-Rehabilitation (REMR). One recent study for the St. Paul District showed that a hole formed by a blown-down tree in the downstream toe area can produce a potentially dangerous increase in hydraulic seepage gradient and internal erosion or piping problems in dikes (Duncan, 1999). The USDA/NRCS referred to the 1950's research work done at the ARS Hydraulics Laboratory in Stillwater, Oklahoma, on Flow in Vegetative Channels, which could have application to some emergency spillways.

A recent literature review, sponsored by ASDSO/FEMA and conducted for the Steering Committee on Plant and Animal Penetration of Earthen Dams, researched available material on the effects of woody plants on dam safety (Tschantz et al, 1999). All types of sources and searches were inventoried, including ASDSO conference and workshop proceedings, ASCE technical journals and articles, USCOLD, direct e-mail and telephone contacts of selected federal and state agency officials, universities, research laboratories and other data bases accessible through the National Technical Information Service (NTIS) and National Performance of Dams Program (NPDP). While only a few references were found on recent or current research of tree and plant effects on dam safety, several references on federal and state practices, policies, and procedures for dealing with trees and vegetation were cited in such topical areas as:

- woody plant physiology
- documented examples of woody plant-caused dam failures, operation, and maintenance problems
- case histories related to tree-caused dam failures
- current and past federal, international, and other research activities
- federal, state, international, and other organizations' policies, procedures and practices for preventing and remediating woody plant problems, and
- federal, state or private cost documentation for removing or controlling trees and woody plants.

Costs of Removing Trees and Tree Related Remediation

Limited cost information for removing trees and brush or for repairing damages caused by vegetation at dams was available from the states or federal agencies. Most state dam safety officials indicated either that they did not have the data or that the owner or his consultant would have that information. Virginia reported that, while costs can be nominal, in extensive tree growth situations where grubbing is required, $10,000 to $20,000 per dam is common and that at one dam; the tree-clearing cost was about $40,000. Missouri reported that such costs could range from $1,000 to $10,000 depending on how badly the dam is overgrown with trees. A prominent North Carolina geotechnical engineering firm stated that ten different contractors, working in North Carolina, South Carolina, and Georgia, reported recent bid prices ranging from about $1500 to $3000 per acre for cutting trees at ground level, removing stumps and root balls, and grubbing the area to remove perimeter roots. Contractors were advised that clearing and grubbing would be done on embankment slopes ranging from 1.5:1 (Horizontal to Vertical) to 4:1 (Horizontal to Vertical), within possible wet areas in the lower 1/3 to 1/2 of the downstream slopes, and on earthen dams ranging in height from 25 to 50 feet. Table 1 compares cost experiences reported by state dam safety officials in different regions of the country for clearing and grubbing trees from dams.

Reporting State	Number of Dams	*Cost per acre	Survey Comments
Georgia	More than 25	$1,000 to $5,000	Based on consultants' feedback; cost varies depending on dam face conditions such as slope steepness, degree of wetness and tree density.
Oklahoma	1 1	$900 $1,150	2 acres of d/s slope over 2-1/2 day period 3-1/2 acres, current proposal estimate.
South Dakota	Several	$100 to $200/Acre	Usually 10 - 20 trees per dam
Nevada	1	$532	Based on 3 hourly laborers working for 2 weeks on 3.25 acres of willow & mesquite removal on d/s dam face (~1995)
Michigan	General DNR construction cost experience	$3,500 $6,000 $12,000	Light clear/grub (diam.<6") Medium clear/grub (diam.<12") Heavy clear/grub (diam<24")
Tennessee	7	$1,540 (Ave.) (approx. range = $1030 to $3290)	Total clearing, grubbing & reseeding cost for 7 dams = $16,705 @ ~1.5 acres per dam. Jobs included range of tree sizes & heavy brush. (1995-98)
Texas	1	$5,500	Part of overall site clearing and grubbing contract for new dam in East Texas (1995)
Ohio	1	$10,000	Cost included clearing, grubbing, mulching and seeding. Heavily wooded; hundreds of trees removed from d/s slope (1999)
Minnesota	Current estimates from Minnesota consultant Small Projects	$1350 $2800 $4475 $4225 $6775 $960	Clearing brush with brush saw - no grubbing Clearing brush by hand - no grubbing Clear and grub brush, incl. stumps Cut & chip up to 6" trees; grub/remove stumps Cut & chip up to 12" trees; grub/remove stumps 16 m-hrs @ $60/hr to clear and grub small trees (diam. < 6") for less than one acre projects

*Reported costs not indexed

Table 1. Cost Comparisons for Clearing, Grubbing and Removing Trees from Dams.

While the range of remedial costs varies widely, depending on several factors, it appears that about $1,000 - $5,000/acre may be a reasonable baseline to use for rough estimating purposes, with the lower figure applicable to small and low-density tree growth and the larger figure appropriate to mature, very dense tree stands.

A typical 25-foot high by 750-foot long earthen dam having 3:1 (Horizontal to Vertical) embankment slopes, a 15-foot crest width, and a freeboard of 10 feet above normal pool has approximately two acres of exposed crest and face area for potential tree growth. Total costs for clearing and grubbing trees for such a dam would be in the range of $2000 to $10,000 depending upon the local site conditions.

Several site-specific factors can influence tree removal costs. These include size and type of trees, growth density, total job size (number of acres), location of growth (crest and/or both faces), embankment slope steepness, slope condition (such as degree of wetness or surface texture), degree and type of required surface treatment (backfilling, use of herbicides or bio-barriers, mulching, seeding, fertilizing, etc.), and regional labor and construction indices.

The U. S. Bureau of Reclamation reported detailed cost data using three herbicidal application methods (aerial, cut-stump, and ground-based foliar-application) in its 1987-93 program to control salt cedar along waterways in seven states of the Upper Colorado Region. Application costs ranged from about $60/acre for aerial spraying to about $1000/acre for cut-stump and spray methods (Sisneros, 1994). The National Park Service indicated that it has done tree removal with the assistance of the U. S. Bureau of Reclamation, but cost information is not readily available.

Summary

Trees appear to be a major dam safety issue for many states. Based on recent survey responses from 48 states, it is estimated that about one half of the state-regulated dams have trees growing on them. The same reporting states estimate that an average of nearly a third of the dams that they regulate have sufficient trees, brush and other growth to hinder effective safety inspections.

Current state and federal policies, procedures, and practices relating to tree and woody plant removal, control, and management for dam safety are generally fragmented and inconsistent among state and federal dam safety agencies. *However, all state and federal agency dam safety officials and experts agree that trees have no place on dams and need to be managed and controlled on both existing and new dams for at least three important reasons:* (1) trees and dense vegetation hinder effective dam inspections; (2) tree roots can cause serious structural instability or hydraulic problems, which could lead to dam failure and possible loss of life; and (3) trees and brush attract burrowing animals, which can in turn cause serious structural or hydraulic problems.

The fragmentation among state and federal agencies applies only to procedures about *how* and *to what extent* the trees and their roots should be removed and resulting cavities remediated to ensure a hydraulically and structurally safe dam. Other chapters in this *Manual* present methods and practices for controlling trees and woody plants and for remediating damage caused by trees and other woody plants.

While limited information is available, a sampling of state dam safety officials and other experts report that the cost of removing trees and brush from the face of a dam can broadly range from about $1,000 to $10,000 per acre, depending on several factors. Typically, the cost of clearing and grubbing trees from dams falls into the $1,000 - $5,000 per acre range. The broad range of costs is not surprising as most dam safety engineers agree that tree removal costs are very much site specific. Controlling vegetation annually is relatively inexpensive, but removing trees on and repairing damage to neglected dams may cost owners several thousand dollars.

Most dam safety experts agree that research needs to be done on determining the relationship of plant and tree species to root penetration of artificial environments such as embankment dams; the interaction between root systems and the phreatic zone and surface; and development and understanding of various types of physical, biological, and chemical treatment and barriers for controlling root growth. Because many existing dams exhibit dense growths of trees and woody vegetation with deep-penetrating root systems, engineering methods need to be developed for understanding, predicting, and stabilizing the effects of these root penetrations to minimize internal erosion and failure. Dam safety experts agree that both technical and non-technical pamphlets and brochures, practice manuals, web-based documents, workshops, and guidance materials need to be developed for educating dam owners about the problems caused by trees and woody vegetation. Engineers, dam safety officials and inspectors, developers, and contractors must be provided technical training and information relative to the control and/or safe removal of trees and other undesirable woody vegetation from earthen dams.

References

1. Association of State Dam Safety Officials (ASDSO), <u>State Survey: Animal and Vegetative Impacts on Dams, Part I - Vegetation on Dams</u> (7 questions), September 1999.

2. Association of State Dam Safety Officials (ASDSO), <u>State Survey, Percentage of Trees on State-regulated Dams</u> (2 questions), January 2000.

3. Soil Conservation Service (SCS), U. S. Department of Agriculture, <u>Technical Note 705 – Operations and Maintenance Alternatives for Removing Trees from Dams</u>, South Technical Center, Fort Worth, April 1, 1981, 8 pages.

4. Tschantz, B. A. and Weaver, J. D., <u>Tree Growth on Earthen Dams: A Survey of State Policy and Practice</u>, University of Tennessee, Civil Engineering Report, November 1988, 36 pages plus Appendices A and B.

5. Tschantz, B. A., Wagner, C. R., Jetton, J. W., and Conley, D. C., <u>Bibliography on the Effects of Woody Vegetation on Dams</u>, compiled for the Association of State Dam Safety Officials (ASDSO) Steering Committee on Plant and Animal Penetration of Earthen Dams, University of Tennessee, September 1999, 18 pages.

6. Tschantz, B. A., <u>Overview of Issues and Policies Involving Woody Plant Penetrations of Earthfilled Dams</u>, Presentation and Proceedings, ASDSO/FEMA Specialty Workshop on Plant and Animal Penetrations on Dams, November 30 - December 3, 1999, 8 pages.

7. Duncan, J. M., <u>Review of Corps of Engineers Design for Rehabilitation of the Perimeter Dikes around Cross Lake, Minnesota</u>, Report submitted to St. Paul District, Corps of Engineers and R. Upton, Ad Hoc Committee Chair, Cross Lake, July 14, 1999, 16 pages plus Appendices A through C.

8. Sisneros, D., <u>Upper Colorado Region Salt cedar Cost Analysis/Evaluation</u>, U. S. Bureau of Reclamation, Research and Laboratory Services Division, Environmental Sciences Section, Denver, Co, Final Report, Memorandum No. 94-2-2, February 1994, 272 pages.

9. U. S. Army Corps of Engineers (USCOE), in cooperation with the Federal Emergency Management Agency (FEMA) and Association of State Dam Safety Officials (ASDSO), <u>National Inventory of Dams - 1998-99</u>, CD-ROM NID-GIS, v. 1.0, with Information Booklet, September 1999.

10. Marks, B. Dan, S&ME Engineering, Inc., Arden, N. C., Faxed communication on recent contractor-bid clearing and grubbing costs, February 23, 2000.

11. Association of State Dam Safety Officials (ASDSO), <u>Report on Specialty Workshop #1: Plant & Animal Impacts on Earthen Dams</u>, Knoxville, Tennessee, Nov. 30 – Dec 3, 1999, June 2000.

Chapter 3
Tree Growth and Tree Root Development Requirements

The purpose of this chapter is to provide the reader and user of this *Manual* with a basic understanding of plant physiology related to fundamental processes of tree growth and tree root development. It is not the intent of this chapter to delve into a detailed biological study of trees and woody vegetation, but to provide the reader with a fundamental understanding of the requirements for tree growth and tree root development while attempting to dispel some of the misconceptions and myths associated with tree and woody vegetation growth, particularly as related to tree root development.

Common Myths and Misconceptions

There are many misconceptions and common myths relating to trees and woody vegetation that have been accepted by many people without a scientific basis. Many of these common myths and misconceptions relative to plant physiology have originated from uneducated interpretations of observations associated with tree growth and tree root development. Some of these myths and misconceptions associated with trees and woody vegetation affect correct interpretation and understanding of the impact of such growth on the safety of earthen dams. The more common myths and misconceptions must be dispelled so that a new level of understanding about the impacts of trees and woody vegetation on earthen dams can be properly developed. Trees and woody vegetation, like all living things, must have oxygen, nutrients, and water (moisture) to survive. Without these requirements, tree roots cannot continue development and tree growth cannot continue. The root system of trees and woody vegetation is in simplified terms comprised of two major components that are the root ball, typically directly below the trunk of the tree, and the lateral or perimeter transport root system that typically extends beyond the 'drip line' or vertical projection of the canopy of the tree.

Tree Tap Roots are thought by many to be the primary root system for all ages and types of trees and woody vegetation. In fact, the taproot is the first root to develop from the seed or reproductive source. This central root is an extension of the stem and differs from the stem only in that the root contains nodes for development of additional roots. Once the taproot has stabilized the young plant (tree), the root ball begins to develop and the taproot becomes less important than other roots that grow laterally from the taproot. The developing root ball provides vertical support for the tree as well as providing nutrients and water (moisture) to the tree. Roots extending laterally from the root ball increase the stability of the tree while functioning to collect and store nutrients, oxygen, and water for the tree. While it is true that some trees have more clearly defined taproots, taproots of most trees do not extend significantly far below the massive root ball of healthy trees. However, taproots are more predominant in locations where trees grow in deep deposits of loose dry soils.

Tree Root Soil Stabilization is likely the most common misconception associated with tree growth and tree root development. How many times has the reader heard, or perhaps mistakenly said, **"If it were not for those trees and tree roots this slope would really be eroded or unstable – those tree roots are really 'holding' that soil slope"**. Many otherwise knowledgeable and educated individuals believe the myth that tree roots actually stabilize soil masses by 'holding' the soil together. This misconception leads many people to believe that heavy tree and woody vegetation growth is actually beneficial for steep embankment slopes. Tree root development that is necessary to provide nutrients for tree growth and stabilize the tree actually loosens the soil mass. Laterally extending tree roots could be thought of as being nature's original application of the geotechnical engineering design concept of soil nailing. *Root penetration stabilizes the tree and loosens the soil mass within which the tree roots are developing; the converse is a myth and certainly not true.*

Groundwater Penetration by tree root systems is another common myth and misconception believed by many otherwise knowledgeable individuals. Although Cypress, Tulip Poplar, some Willow and Water Birch tree species appear to have root systems that are submerged, nutrient root systems of trees cannot survive beneath the water table or the phreatic surface (seepage line) in an earthen dam. Trees and woody vegetation depend upon their transport root systems to provide the major portion of the oxygen demand for continual tree growth and tree root development. Most species of trees and woody vegetation quickly die of suffocation once the lateral transport root system and root ball are inundated. This phenomenon can be visually observed in many areas of Arkansas, Mississippi, and Louisiana where large tracts of timber have been artificially flooded for duck hunting. If these flooded tracts of timber are not drained seasonally, the timber (trees) die as a result of suffocation. Similarly, beaver activity causes significant losses in the timber industry every year as a result of inundation of harvestable timber. Tree roots do not penetrate the water table or the zone of saturation where oxygen demands of the tree cannot be met. If the zone of saturation or water table is raised above the level of tree roots for an extended period, the tree will die as a result of suffocation. Tree root development and tree growth cannot occur when moisture contents in the soil mass are greater than about forty percent.

Soil Moisture Uptake of many species of trees far exceeds that which most individuals would estimate as a normal requirement of water for continual tree growth and tree root development. It is not uncommon for most species of healthy mature trees to absorb 200 to 300 gallons of water per day if this amount of water is available to the lateral transport root system. Reduced availability of soil moisture will curtail continual tree root development until such time that the soil mass is replenished with sufficient moisture to allow resumption of tree root development. Tree root development and tree growth cannot occur in soil masses having moisture contents less than about twelve percent for extended periods.

Woody Vegetation Control Versus Dam Performance is an issue that is clearly misunderstood by many dam owners, operators, inspectors, dam safety regulators, engineers, and consultants. Tree and woody vegetation root penetration is not a beneficial effect on the performance of earthen dams. As indicated previously, tree root penetration does not stabilize a soil mass, particularly an embankment slope. Quite the contrary, tree root penetration loosens the soil of an embankment slope and creates a condition more conducive to surface water penetration and slope failure. Earthen dams are not unlike other engineered structures in that they must be properly maintained in order to perform as perceived in the original design of the structure.

When does routine vegetation maintenance and control become a dam safety and/or dam performance issue? The author is of the opinion that vegetation maintenance and control on an earthen dam ends, and the need for an *engineered* earthen dam rehabilitation plan begins, when effects of an improper vegetation maintenance and control program create conditions that are *detrimental to the structural integrity* of the earthen dam. For example, an earthen dam that exhibits a dense growth of grasses and weeds that are waist high, but is free of significant woody vegetation growth, is an earthen dam that is in need of proper vegetation maintenance and control to allow proper inspection of the dam. However, waist-high grasses and weeds would not typically affect the structural integrity of the earthen dam. Conversely, an earthen dam that supports a dense growth of four to eight inch diameter trees that preclude proper access for inspection is a dam safety and performance issue. Dense growths of trees and woody vegetation not only present a hindrance to proper dam safety inspection, but also are detrimental to the structural integrity of the earthen dam. Proper removal of trees and woody vegetation from earthen dams is a dam safety and performance issue that must be conducted in accordance with properly designed dam remediation plans and specifications.

Tree Root Characteristics and Requirements

As previously indicated, root systems of trees and woody vegetation consist of two primary components that are the root ball and the lateral transport root system. While all tree and woody vegetation roots have a primary function of providing oxygen, nutrients, and water to the plant, they also provide stability for the plant. The root ball that is typically directly below the trunk of the tree provides vertical support while the lateral transport roots provide lateral support for the tree. Root systems of trees and woody vegetation growing on dam embankment slopes will typically be asymmetrical as a result of the need for the tree to be stabilized in the sloping embankment soil mass. The lateral transport roots will typically be better developed on the uphill side of the tree than on the downhill side of the tree. Dr. Kim D. Coder at the University of Georgia has conducted extensive studies and research on tree growth and tree root development requirements and characteristics. He has developed data from these studies and research programs that relate tree trunk size to root ball diameter and lateral transport root system diameter. These data are presented in Table 1 below.

Table 1: Typical Rootball and Root System Sizes for Various Tree Sizes

Tree Diameter, inches	Rootball Diameter, feet	Root System Diameter, feet
4 to 5	6	10 to 12
6 to 7	8	16 to 18
8 to 9	10	20 to22
10 to 11	12	26 to 28
12 to 14	14	30 to 32
15 to 18	16	38 to 46
19 to 23	18	48 to 58
24 to 36	20	60 to 90
37 to 45	22	92 to 112

During the presentation of common myths and misconceptions about tree growth and tree
root development, requirements of trees and woody vegetation for continual growth and
root development were discussed. Based upon research and studies conducted by Dr. Kim
Coder, requirements for tree and woody vegetation growth and root development are
tabulated in Table 2.

Table 2: Root Growth Resource Requirements

Requirement	Minimum Value	Maximum Value
Soil Oxygen Content	2.5%	21.0%
Soil Air Voids	12.0%	N/A
Soil Bulk Density (Clays)	N/A	87 pcf
Soil Bulk Density (Sands)	N/A	112 pcf
Water Content of Soil	12.0%	40.0%
Limiting Soil Temperatures	40°F	94°F
Soil pH Values	3.5	8.2

Soil air void content is one of the most critical factors for continual tree root
development. This factor is critical since both soil density and soil oxygen content are
dependent upon the amount of air voids present in a soil mass. Because of the importance
of soil air void content, Dr. Coder conducted extensive research to determine limiting air
void contents for various soil types required for continual tree root growth (See Table 3).

Table 3: Limiting Soil Air Voids for Root Growth in Various Soil Types/Textures

Soil Type/Texture	Air Voids, %
Sand	24
Fine Sand	21
Sandy Loam	19
Fine Sandy Loam	15
Loam	14
Silt Loam	17
Clay Loam	11
Clay	13

Utilizing weight-volume relationships for various soil types and textures, Dr. Coder was able to determine the limiting (maximum) dry density of soil that would allow continual tree root development. Results of these correlations between minimum soil air void content and maximum soil dry densities required for continual tree root development are presented in Table 4 below.

Table 4: Limiting Soil Dry Density for Root Growth in Various Soil Types/Textures

Soil Type/Texture	Dry Density, pcf
Sand	112.3
Fine Sand	109.2
Sandy Loam	106.1
Fine Sandy Loam	103.0
Loam	96.7
Silt Loam	90.5
Clay Loam	93.6
Clay	87.4

In an attempt to relate the research data developed by Dr. Coder to geotechnical engineering data developed from over 200 earthen dam projects, the author has compiled a comparative list of soil properties for various soils that have been found in earthen dam embankments. The ranges given in the data presented in Table 5 below are associated with soil in a loose condition and soil in a compacted state that might be required in the construction or remediation of an earthen dam. *The user of this Manual must be aware that these soil parameters are typical values and should not be relied upon for design of new earthen dams or design of remediation plans for existing dams.*

Table 5: Summary of Typical Soil Parameters

Soil Type	Specific Gravity	Void Ratio	Porosity, %	Dry Density, pcf	Permeability, cm/sec
Sand	2.62 to 2.66	0.40 to 0.90	30 to 45	90 to 115	0.01 to 0.0001
Silt	2.60 to 2.68	0.50 to 1.20	35 to 55	75 to 110	0.001 to 0.00001
Clay	2.66 to 2.72	0.60 to 1.40	40 to 60	70 to 105	0.0001 to 0.0000001

As one can see from the tabulated summary of typical soil parameters, continual tree root development cannot occur in soils that are well compacted. One of the best methods of controlling tree and woody vegetation growth on new earthen dams and existing earthen dams where remediation requires placement of additional embankment fill soil is to compact the embankment fill soils to a high degree of compaction. Increased compaction of embankment fill soils reduces the air void content and limits the amount of surface water that can infiltrate into the embankment slope. However, a good ground cover of grasses can be established in well-compacted soils since the depth of grass root penetration is minimal and the surficial soils will typically sustain the shallow grass root penetration.

References

1. Association of State Dam Safety Officials (ASDSO), <u>Report on Specialty Workshop #1: Plant & Animal Impacts on Earthen Dams</u>, Knoxville, Tennessee, November 30 – December 2, 1999, June 2000.

2. Coder, K. D., <u>Tree Root Growth Control Series: Root Growth Requirements and Limitations,</u> Univ. of Georgia, Cooperative Extension Service Forest Resources, Publication FOR98-9, March 1998, 8 pp.

3. Coder, K. D., <u>Tree Root Growth Control Series: Soil Constraints on Root Growth,</u> Univ. of Georgia, Cooperative Extension Service Forest Resources, Publication FOR98-10, March 1998, 8 pp.

4. Coder, K. D., <u>Engineered to Fail? Tree Root Management on Dams</u>, Abstract, University of Georgia, Athens, November 1999, 1 page.

5. Dickerson, William C., <u>Integrative Plant Anatomy,</u> Harcourt Academic Press, New York, 2000

Chapter 4
Earthen Dam Safety Inspection and Evaluation Methodology

The purpose of this chapter is to illustrate dam behavior during the initial years of design life and to present a suggested inspection and evaluation methodology. An example earthen dam configuration will be presented in order to illustrate earthen dam behavior and to develop the suggested inspection methodology.

Example Earthen Dam Configuration

The example earthen dam is assumed to be a high-hazard dam having a structural height of about 33 feet and impounding a lake area of about three acres at normal pool elevation. The contributing watershed of the lake is about 320 acres (0.5 square mile) with a base flow of about one-half (0.5) cubic feet per second (cfs).

The configuration of the example earthen dam consists of an upstream slope of 2:1 (horizontal to vertical), a crest width of fifteen feet, and a 3:1 (horizontal to vertical) downstream slope. The dam has a freeboard of four feet making the hydraulic height of the dam about 29 feet. The dam is founded on relatively impervious (compared to the embankment fill soil) material with a down gradient slope of about three percent. The example earthen dam section has a key trench directly below the centerline of the dam crest that has a bottom width of ten feet and side slopes of 1:1 (horizontal to vertical). The dam crest has a two-percent slope toward the impounded lake and the upstream slope has no protection system against tree and woody vegetation growth or wave erosion. The embankment of the example earthen dam is assumed to be homogeneous. Figure 1 is a representation of the example earthen dam configuration with the theoretical seepage line intercepting the downstream slope at about one-third the hydraulic height of the dam. **Rule-of-Thumb:** *The phreatic surface intercepts the downstream slope of a homogeneous earthen dam at a vertical distance of about one-third the hydraulic height above the toe of the downstream slope, provided there is no internal drainage system in the dam embankment.*

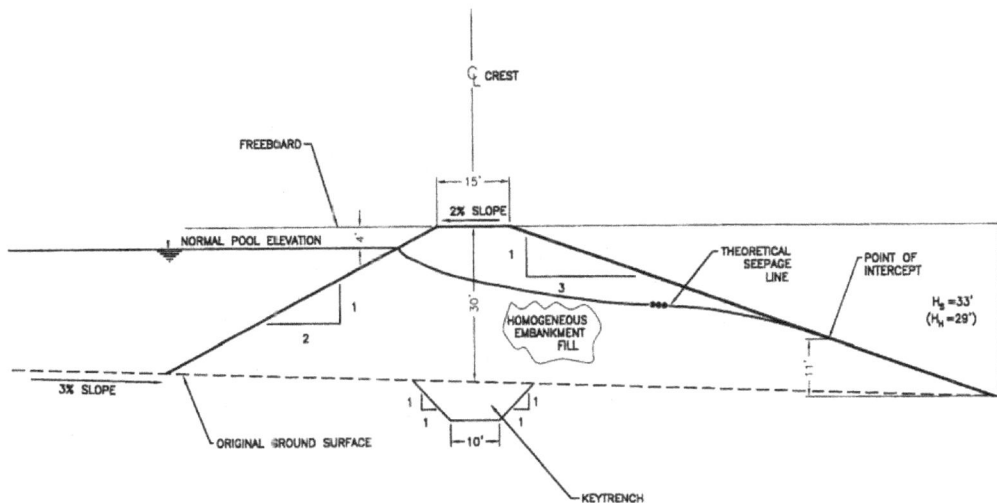

TYPICAL EMBANKMENT SECTION
WITHOUT TOE DRAIN SYSTEM

Figure 1

Based upon data provided for the example earthen dam, this dam would be listed on the National Inventory of Dams (NID). In addition, the example earthen dam would be classified as a small-size, high-hazard dam by most state dam safety regulations.

Figure 2 illustrates the example earthen dam with an embankment subdrain system located within the downstream embankment slope. The subdrain or embankment drain system is located at about the point of interception of the seepage line with the downstream slope if there was no embankment toe drain system within the downstream slope. As a result of the presence of the embankment subdrain system, the seepage line through the dam embankment has been modified (lowered) from the location of the theoretical seepage line for a homogeneous earthen dam embankment. The seepage line within an earthen dam is often mistakenly considered to have a permanent location.

However, the location of the seepage line is continually changing as a result of many influential factors. Fluctuations in the pool elevation, seasonal and long-term climatological conditions, and the growth of trees and woody vegetation in close proximity to the seepage line are some of the factors that influence changes in the location of the phreatic surface within an earthen dam embankment.

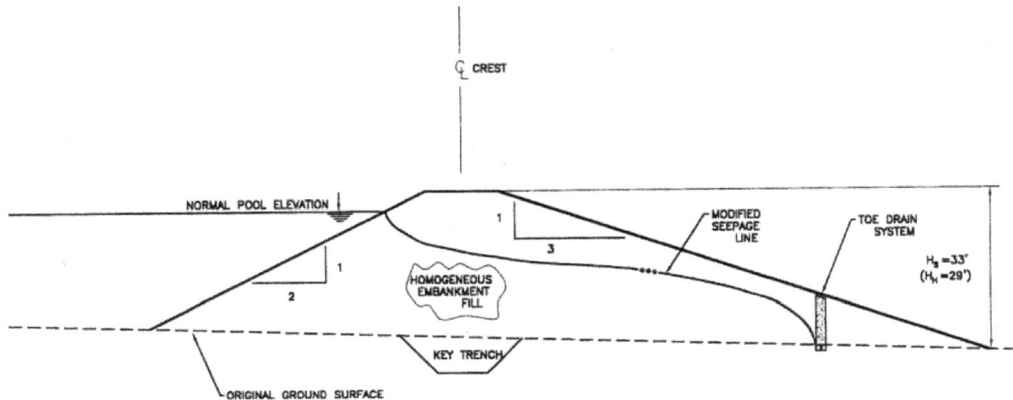

**TYPICAL EMBANKMENT SECTION
WITH TOE DRAIN SYSTEM**

Figure 2

Important moisture regimes other than the steady-state seepage line (phreatic surface) are often not given proper consideration in the evaluation of the performance of earthen dams. The *zone of saturation* is located immediately above the phreatic surface or seepage line where embankment fill soils have become saturated as a result of capillary rise caused by capillary forces in the soil voids. Figure 3 illustrates the presence of zones of saturation associated with that of a theoretical seepage line location without an embankment subdrain system as well as that of a modified seepage line location with an embankment toe drain system. The height of capillary rise (thickness of the zone of saturation) is directly dependent upon the effective mean diameter of soil voids within the earthen dam embankment. The effective mean diameter of compacted soil is dependent upon the *effective particle size* (De) of the compacted embankment fill soil. Soil within the zone of saturation is completely saturated; however, there is no flow or gravity induced movement of water unless some external force disturbs the soil. This phenomenon is often observable during the inspection of downstream slopes of earthen

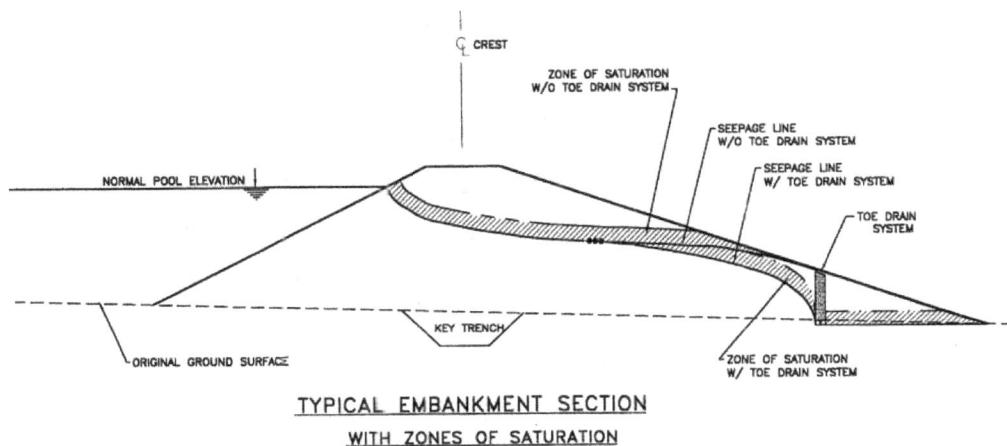

TYPICAL EMBANKMENT SECTION
WITH ZONES OF SATURATION

Figure 3

dams. Seepage and free flowing water can be seen on the downstream slope of an older dam below the point of interception of the seepage line if no embankment subdrain is present. Above the point of interception of the seepage line with the downstream slope, the soil is saturated and the Zone of Saturation can be observed for a significant distance above the seepage line intercept in some cases. In the Zone of Saturation, pore water may be observed to fill tracks made in the water-softened embankment soil. However, once the tracks are filled by pore water released from the disturbed soil there will be no continued flow or seepage from the embankment. This condition is often confused with the presence of embankment seepage. Installation of a subdrain location in this situation may lower the phreatic surface relatively quickly; however, months or even years may be required to drain the zone of saturation because of tensile forces or negative pore pressures in the embankment fill soils.

Embankment Wetting, Saturation, and Seepage

Prior to presentation of the behavior and performance of an earthen dam embankment during the initial years of the design life, one must have an understanding of relationships between various velocities of moisture movement and water flow through compacted embankment soils. First, consider the relationship between the optimum compaction moisture content of an embankment soil and other moisture content properties.

Rule-of-Thumb: *The optimum compaction moisture content as determined by ASTM D-698 (standard Proctor compaction test) is approximately two to four percent below the Plastic Limit (PL) of most soils and about three to five percent below the saturation moisture content of the same soils.*

Compacted soils will typically increase in moisture content from the compaction moisture content to about the PL of the soil relatively quickly after construction of an earthen embankment. The rate of wetting is much greater in soils compacted dry of optimum moisture content than in soils compacted wet of optimum moisture content. Although compacted soils may undergo wetting or increase in moisture content relatively quickly when exposed to a source of water, the rate of saturation is much slower because air trapped in discontinuous soil voids must be dissolved in soil pore water during the saturation process. Embankment wetting and saturation are not associated with seepage or the flow of water through a homogeneous earthen dam; however, relative velocities of wetting and saturation can be related to values of steady-state seepage velocity, permeability, or hydraulic conductivity of compacted embankment soils.

Figure 4 is an illustration of the example earthen dam with relationships between various soil water flow velocities and permeabilities. First, consider the relationship between the vertical and horizontal permeability of a compacted homogeneous embankment soil.

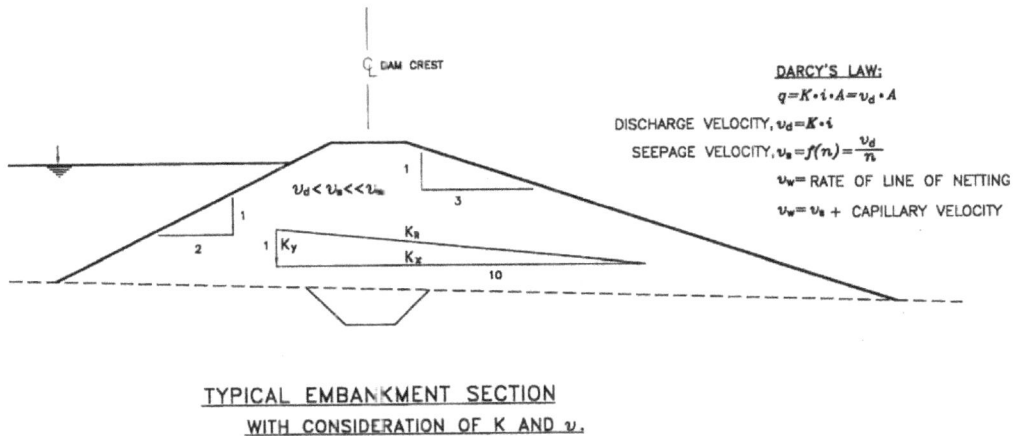

TYPICAL EMBANKMENT SECTION
WITH CONSIDERATION OF K AND v.

Figure 4

Rule-of-Thumb: *The horizontal permeability of a compacted homogeneous embankment soils are typically about nine times to ten times (one order of magnitude) greater than the vertical permeability.*

Variation between the horizontal permeability and vertical permeability is the result of the internal structure of compacted soils. This variation does not account for poorly compacted lifts since the embankment is assumed to be homogeneous. Consequently, if laboratory permeability tests indicate that a compacted embankment soil exhibits a hydraulic conductivity value of about 0.000004 centimeters per second (cm/sec) then the horizontal permeability of this compacted embankment soil will be about 0.000036 to 0.00004 cm/sec. Second, consider Darcy's Law that is the basis for all theories and analyses associated with the flow of water through soil masses. Darcy did not account for soil voids relative to soil solids in derivation of his equation. As a result, the area of discharge is the total cross-sectional area through which flow is occurring. If one assumes that the hydraulic gradient producing flow through a soil mass is equal to one (unity), then the ***discharge velocity*** (Darcy's flow velocity) is equal to the permeability value of the soil. The actual flow velocity in the voids of the soil is often identified as the *seepage velocity* and is approximately equal to the discharge velocity divided by the porosity value (expressed as a decimal) of the soil. Assuming that the compacted embankment soil in the example earthen dam has a porosity of forty (40) percent (0.40), the seepage velocity of the soil would be about 2.5 times greater than the discharge velocity. Third, consider the wetting velocity or the velocity of the *line of wetting*. The wetting velocity is the rate at which soil increases in moisture content up to about the PL when exposed to a free water source. The line of wetting can often be observed as it progresses through soil masses, particularly soils that are dry of optimum moisture content. The wetting velocity is the sum of the seepage velocity and the capillary velocity or the velocity of wetting attributable to capillary forces in the soil.

Rule-of-Thumb: *The wetting velocity or the velocity of the line of wetting through compacted soil is about one order of magnitude (ten times) greater than the seepage velocity.*

Applying this factor to the previous comparison between seepage velocity and discharge velocity, one finds that the wetting velocity is about 25 times greater than the discharge velocity. Based upon the foregoing discussion of earthen dam embankment wetting, saturation, and steady-state seepage velocities, consider the illustration in Figure 5. This figure illustrates embankment wetting, saturation, and steady-state seepage during the early years of the design life of an earthen dam. Assume that laboratory testing indicates that embankment soils of the example dam embankment have a permeability or hydraulic conductivity value of 0.02 foot per day. The discharge velocity would be about 0.008 foot per day with a hydraulic gradient of about 0.4 resulting in a horizontal discharge velocity of about 0.08 foot per day. The associated seepage velocity would be about 0.02 foot per day with a soil porosity of about 40 percent and the horizontal seepage velocity would be about 0.2 foot per day. The velocity of the line of wetting or the wetting velocity would be about 2.0 feet per day.

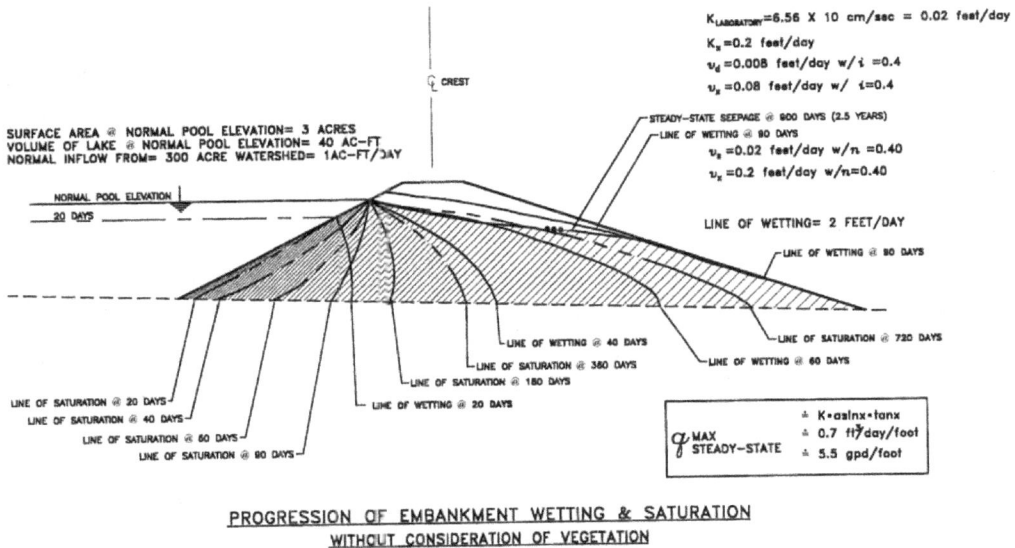

PROGRESSION OF EMBANKMENT WETTING & SATURATION
WITHOUT CONSIDERATION OF VEGETATION

Figure 5

Based upon the estimated normal inflow from the contributing watershed, the lake retained by the example dam should reach about fifty percent volume in approximately twenty days and reach normal pool elevation in about forty days. Solid lines in Figure 5 illustrate the location of the line of wetting at various time intervals. The line of wetting should reach the downstream slope in about ninety days. *Note: The compacted embankment soils remain partially saturated after passage of the line of wetting.* Dashed lines in Figure 5 illustrate the line of saturation at various time intervals. The line of saturation moves at the seepage velocity that is about one-tenth the value of the wetting velocity. When the line of wetting has reached the downstream slope in about ninety days, the line of saturation is still at about the vertical from the intercept of the normal pool with the upstream slope. Based upon this rate of progression, the line of saturation will not reach the surface of the downstream slope and steady-state seepage will not be initiated for about 900 days (about 2.5 years), *provided that no external influences affect the rate of wetting and saturation*.

The estimated maximum steady-state seepage rate for the example dam will be about 5.5 gallons per day per foot of dam. Before leaving Figure 5, imagine that the example dam contains an embankment subdrain system as indicated in Figures 2 and 3. The rate of progression of the line of wetting and the line of saturation will both be affected by the presence of the subdrain system.

Even without the presence of an embankment subdrain system, the time required for the line of wetting could encompass an entire growing season depending upon the time of year that the dam was completed. More importantly, the time that is required for the line of saturation to intercept the downstream slope might encompass two or three entire growing seasons. Tree and woody vegetation growth can become quite dense and relatively large within the initial two to three growing seasons if not properly controlled.

The initiation of tree and woody vegetation growth on the downstream slope begins the soil moisture uptake cycle so that the line of saturation and the seepage line may never completely develop and intercept the downstream slope. The condition represented by Figure 6 might initially be considered to be beneficial to the stability of the dam embankment. However, one must understand that as the tree and woody vegetation growth continues compacted soils of the dam embankment are continually loosened by the penetration of major tree and woody vegetation root systems. Furthermore, trees that might appear healthy to an untrained inspector may be an unhealthy specimen and have a premature death leaving penetrating root systems to rot inside the dam embankment. Additionally, soil nutrients in the compacted soil embankment of an earthen dam may not be sufficient for development of growth beyond which the tree cannot be properly sustained without premature death. Regardless of the cause, trees and woody vegetation do die and cease to uptake soil moisture that they previously used. This change in soil moisture uptake will affect the zone of aeration, zone of saturation, and the location of the seepage line in the vicinity of the unhealthy or dead trees and woody vegetation.

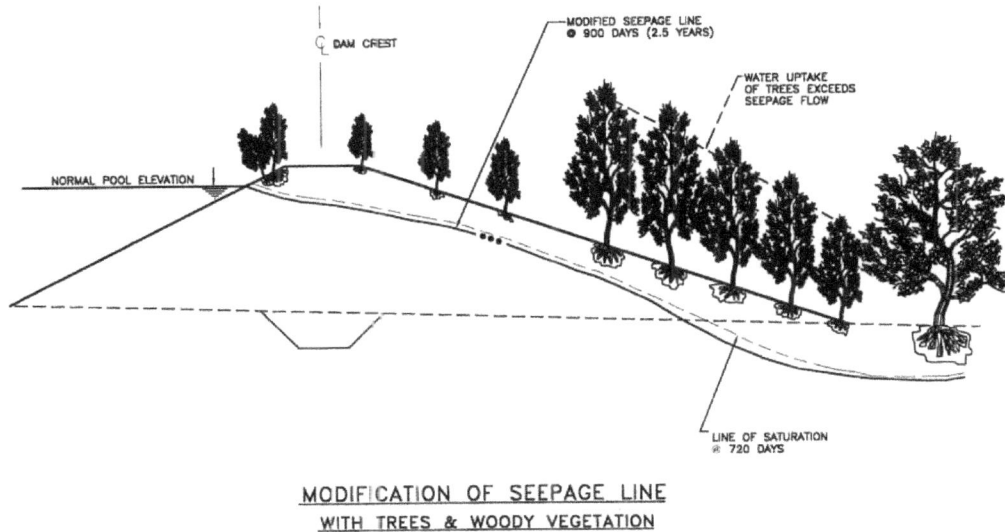

MODIFICATION OF SEEPAGE LINE
WITH TREES & WOODY VEGETATION

Figure 6

The Mid-Life Crisis of an Aging Earthen Dam

Once an earthen dam embankment has become impregnated with numerous trees and woody vegetation penetrations, routine and even major maintenance activities will likely not be sufficient to regain the original design life of the dam. At this time in the life of an earthen dam, previously identifiable maintenance problems have become serious dam performance and dam safety issues. Restoration through an ***engineered*** dam remediation design and remediation construction is typically required to bring the dam to acceptable standards relative to dam safety requirements.

Figure 7 illustrates some of the problems and dam safety issues that can be created by uncontrolled or non-maintained tree and woody vegetation growth in what has been termed by the author as the *'Mid-Life Crisis'* of an earthen dam. Seepage flow may be emerging from rootball cavities of blowdowns (uprooted trees) because they are no longer using soil moisture and the seepage line has adjusted upward toward the surface of

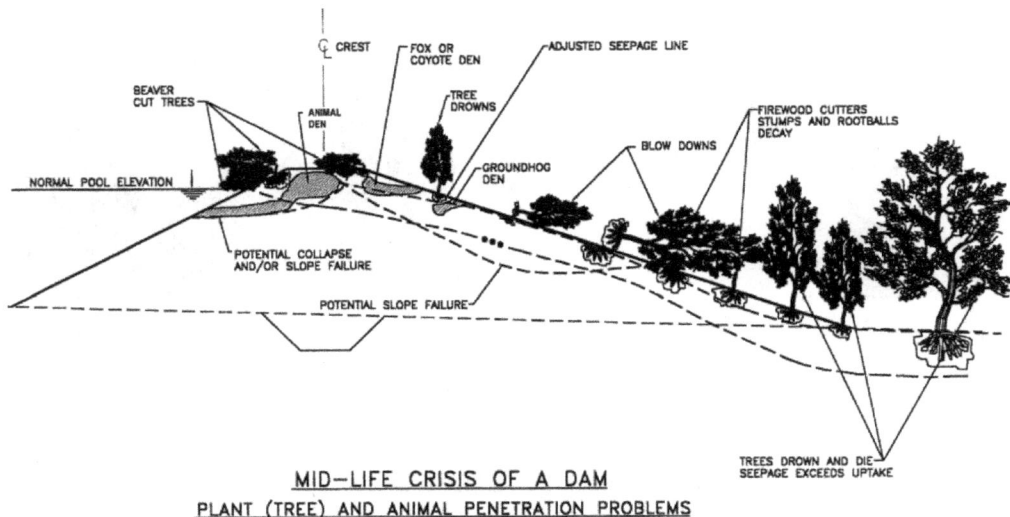

MID—LIFE CRISIS OF A DAM
PLANT (TREE) AND ANIMAL PENETRATION PROBLEMS

Figure 7

the slope. Removal of mature trees by woodcutters deletes the soil moisture uptake of the removed trees thus further modifying the location of the seepage line closer to the surface of the downstream slope. Rootballs and root systems of otherwise healthy trees located at and beyond the toe of the downstream embankment slope become inundated by the adjusted seepage line. Since trees cannot live through prolonged submergence of their major root systems, these trees will become unhealthy and die leaving decaying rootballs and root systems as serious penetrations in the earthen dam. Rootball cavities remaining from blowdowns (uprooted trees) and their relationship to the seepage line create conditions susceptible to potential slope failure of the downstream embankment slope. Restoration of the example earthen dam illustrated in Figure 7 to a safe condition cannot be brought about through routine maintenance activities. An ***engineered*** dam remediation design and remediation construction will be required to restore this dam to a safe condition and original design life.

Inspection and Evaluation Methodology

The effectiveness, economics, and constructability of dam remediation designs for earthen dams begin and end with proper evaluations of the characteristics and seriousness of deficiencies as related to dam safety issues. ***All tree and woody vegetation growth on earthen dams is undesirable and has some level of detrimental impact upon operation, performance, and safety of an earthen dam.*** However, not all tree and woody vegetation growth on earthen dams imposes the same level of impact on operation, performance, and dam safety. Dam owners, regulators, inspectors, and engineers must develop an understanding of the impact of tree and woody vegetation growth relative to location on the dam configuration. Proper evaluation of the seriousness of dam safety issues related to tree and woody vegetation growth on earthen dams is typically associated with the location of the undesirable plant growth on the dam embankment.

A few examples of the variability of seriousness of plant penetrations are presented herein to begin the learning process. The presence of a twelve-inch diameter tree on the downstream side of the crest of an earthen dam typically does not pose the same degree of impact on potential dam safety as a twelve-inch diameter tree located in the lower portion of the downstream slope. Conversely, a twelve-inch diameter tree in the upper portion of the downstream slope does not typically create the same level of seriousness as an unhealthy twelve-inch diameter tree on the upstream slope or front crest of a dam having a narrow crest width. Ornamental shrubs having shallow root systems along a wide roadway crossing the crest of an earthen dam will not impose the same level of seriousness as similar shallow rooted woody vegetation growing on the lower portion of the downstream slope.

The purpose of developing a well-defined inspection and evaluation methodology is to allow the establishment of dam remediation design priorities. Most anyone having a basic understanding of the seriousness of tree and woody vegetation growth to the safety of earthen dams can inspect an earthen dam and recommend removal of all trees, stumps, and root systems. However, inspectors and dam engineers must develop a definitive inspection and evaluation methodology in order to prioritize the seriousness of various locations of tree and woody vegetation growth on earthen dams.

Many individual dam owners do not have economic resources to undertake extensive dam remediation projects to bring an earthen dam into safe operation and performance conditions if the dam exists in a severely deteriorated condition. These owners often have to budget dam remediation projects over a scheduled maintenance and remediation construction period. Dam safety regulators, inspectors, and engineers that have developed and utilized a well-defined dam safety inspection and evaluation methodology can communicate priorities to dam owners so that the needed dam remediation design components can be completed in a prioritized manner. All too often dam safety regulators and engineers overwhelm dam owners with dam deficiencies without consideration of prioritization of deficiencies on dam safety, performance, and operation.

Dam Safety Inspection and Evaluation Zones

Five dam safety inspection and evaluation zones have been identified within the geometric configuration of a typical earthen dam. The delineated zones, illustrated in Figure 8, are not numbered in any implied order of seriousness relative to the impact of tree and woody vegetation growth, but have simply been numbered from upstream to downstream. The seriousness and potential impacts of tree and woody vegetation growth within each inspection and evaluation zone will be discussed during the description and identification of the delineated dam safety inspection and evaluation zones.

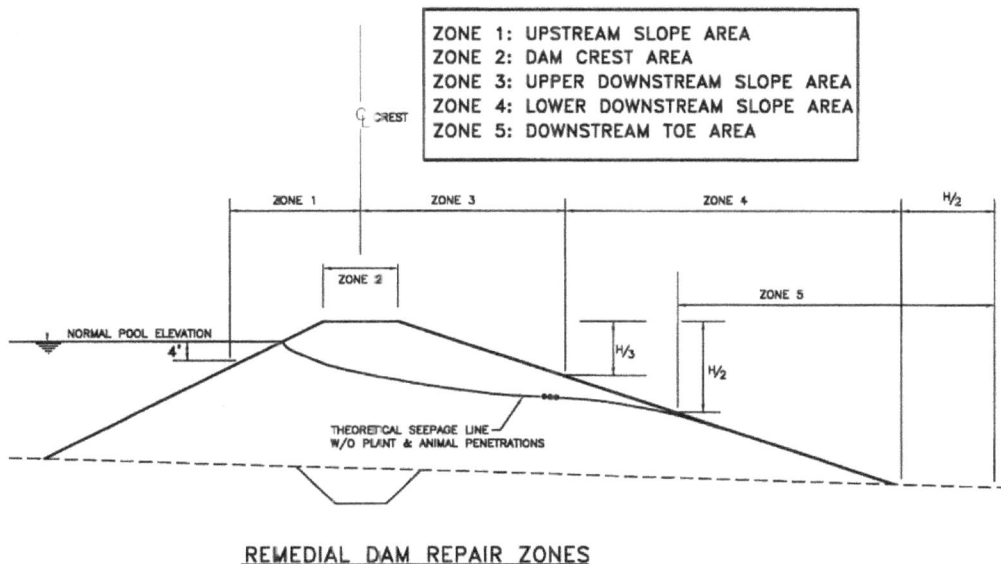

REMEDIAL DAM REPAIR ZONES

Figure 8

Inspection and Evaluation Zone 1 begins on the upstream slope of the earthen dam embankment at about four feet below normal pool elevation. Zone 1 extends laterally to the centerline of the crest of the dam. Tree and woody vegetation growth in Zone 1 is more critical relative to dam safety in the case of dams having a narrow crest width than those having a wide crest width. Zone 1 also includes the area subject to damage resulting from wave erosion and frequently recurring rapid drawdown events.

Inspection and Evaluation Zone 2 includes the entire width of the crest of the dam. Zone 2 overlaps Zone 1 by one-half the crest width. Overlapping a portion of Zone 1 with a portion of Zone 2 was done to emphasize the critical portions of both zones. Zone 2 is typically considered to be one of the least critical zones relative to dam safety issues associated with tree and woody vegetation growth. However, careful inspection of Zone 2 often reveals evidence of serious dam safety issues such as tension cracks, slope failure scarps, and erosion features that may or may not be related to tree and woody vegetation growth originating in other dam safety inspection and evaluation zones.

Inspection and Evaluation Zone 3 extends from the centerline of the crest of the dam to a point on the downstream embankment slope that is about one-third of the structural height below the crest of the dam. Zone 3 overlaps Zone 2 by one-half the crest width and is typically considered the least critical zone relative to dam safety issues associated with tree and woody vegetation growth. The seepage line and zone of saturation in this portion of an earthen dam embankment are typically sufficiently far below the surface to allow excavation of tree rootballs on the downstream slope of the dam without installation of a drain or filter system. A portion of Zone 2 has been overlapped by Zone 3 to draw attention to the most critical portion of Zone 3 that is the downstream portion of the crest of an earthen dam.

Inspection and Evaluation Zone 4 extends from a point on the downstream embankment slope that is about one-third the structural height of the embankment to the toe of the downstream embankment slope. Zone 4 is one of the two most critical zones relative to dam safety issues associated with tree and woody vegetation growth as well as other potential dam safety issues. This zone typically contains the interceptions of both the zone of saturation and the seepage line with the downstream slope. The close proximity of the zone of saturation and seepage line to the surface of the downstream embankment slope in this zone is a critical factor relative to dam safety issues associated with tree and woody vegetation growth. *Tree and woody vegetation growth in this Zone 4 must be of major concern to everyone associated with the safety of an earthen dam and must be evaluated carefully relative to prioritization of dam remediation requirements.*

Inspection and Evaluation Zone 5 extends from the mid-height of the downstream embankment slope to a distance of one-half the structural height beyond the toe of the downstream embankment slope. This zone typically contains the interception of the seepage line with the downstream embankment slope and potential boiling (soil piping) action beyond the toe of the downstream embankment slope. As such, this zone is critical relative to long-term, steady-state seepage stability considerations for an earthen dam. Tree and woody vegetation growth in this zone rapidly develops into serious conditions that directly affect the safety of an earthen dam. Zone 5 overlaps Zone 4 to draw attention to the more critical portions of both Zone 4 and Zone 5. As in the case of Zone 4, Zone 5 is typically considered to be one of the two most critical zones relative to dam safety issues associated with tree and woody vegetation growth. Tree and woody vegetation growth in Zone 5 must be a concern to all involved in the safety of an earthen dam. ***Maintenance and/or engineered dam remediation must be undertaken immediately in the event that tree and woody vegetation growth is significant within Zone 5.*** Control of tree and woody vegetation growth well beyond the toe of the downstream embankment slope cannot be over-emphasized. This area of an earthen dam is critical to overall stability and potential dam safety issues associated with embankment and foundation seepage.

The dam safety inspection and evaluation methodology set forth herein can be easily modified and/or extended to meet the needs of specific dam owners, dam safety regulators and inspectors, and engineers. This proposed methodology for dam safety inspections and evaluations should provide a basic plan that will allow the reader to customize and/or improve existing dam safety inspection and evaluation programs.

References

1. Casagrande, Arthur, "Seepage Through Dams", Contributions to Soil Mechanics: 1925 – 1940, Boston Society of Civil Engineers, pp 295-336 (Originally Published in the Journal of New England Water Works Association, Volume LI, No.2, June 1937.

2. Cedegren, Harry R., Seepage, Drainage, and Flow Nets, John Wiley & Sons, New York, 1967.

3. Marks, B. Dan, "The Behavior of Aggregate and Fabric Filters in Subdrain Applications, Research Report, Department of Civil Engineering, University of Tennessee, Knoxville, Tennessee, February 1975.

4. Means, R. E., and Parcher, J. V., Physical Properties of Soils, Charles E. Merrill Publishing Company, Columbus, Ohio, 1963.

5. Parcher, J. V., and Means, R. E., Soil Mechanics and Foundations, Charles E. Merrill Publishing Company, Columbus, Ohio, 1968.

6. United States Department of Agriculture (USDA), Natural Resources Conservation Services (NRCS), (formerly Soil Conservation Service, SCS), Technical Note 705 – Operations and Maintenance Alternatives for Removing Trees from Dams, South Technical Center, Fort Worth, Texas, April 1981.

7. United States Department of Agriculture (USDA), Natural Resources Conservation Services (NRCS), (formerly Soil Conservation Service, SCS), Technical Engineering Notes – OK-08 (Revised) RE: Control of Trees and Brush on Dams, Oklahoma State Office, Stillwater, Oklahoma, February 1990.

Chapter 5
Controlling Trees and Woody Vegetation on Earthen Dams

The establishment and control of proper vegetation on an earthen dam are essential to maintaining a *safe* dam. Effective, shallow-rooted, vegetative cover is necessary to reduce and prevent embankment slope erosion. Trees and other undesirable deep-rooted vegetation should be prevented from being established for the following reasons:

- Permit effective inspection and monitoring of embankment crest and faces
- Allow for adequate access to dam for normal and emergency operation
- Prevent structural damage from embankment piping and internal erosion, unstable slopes from toppled trees, concrete wall/slab joint cracking/displacement, and other problems
- Reduce possibility of root-blocked drains
- Prevent blockage of spillway channel
- Discourage rodent and other animal activity by eliminating food source and habitat
- Eliminate expensive tree and brush removal and remediation costs
- Reduce impression of owner neglect

Consequently, dam owners should observe these four important rules:

1. Existing trees should be removed and not be allowed to mature on earthen dams, abutment groins, or around water conveyance structures
2. Trees or shrubbery should never be planted on or around new or existing dams
3. Existing trees should be watched closely until they are removed
4. Grasses and shallow-rooted native vegetation are the most desirable surface covering for an earthen dam.

Dam owners should be especially aware of dangerous or potentially hazardous tree conditions such as decaying or dead branches; lightening-caused splits; stripping or breakage; leaning, uprooted or blown-down trees; and seepage around exposed tree roots located along embankment slopes, especially in vulnerable downstream toe or abutment areas. Outward leaning trees may result from a slumping embankment condition that can be an indicator of slope instability. Any of these conditions warrants immediate attention by the owner and a qualified engineer.

Woody vegetation and tree growth creating undesirable root penetrations in earthen dams can be controlled or prevented by proper management of root growth into new dams and dams that have previously been cleared of trees by proper removal procedures. In this manual, some of the characteristics of woody vegetation and tree growth are presented relative to the aging of an earthen dam. Remedial dam repair design procedures and construction techniques are presented for proper removal of trees of various sizes in various areas of the geometric configuration of an earthen dam. Proper management and control of woody vegetation on new and previously repaired dams (tree removal projects) are based upon an understanding of soil conditions that limit root growth, factors that affect or promote root growth, and various procedures and techniques that can be used to stop, redirect, and/or reduce the rate of root elongation.

The purpose of this chapter is to provide a basic understanding of requirements for healthy root elongation, and to provide an introduction to some techniques and procedures that can be utilized to manage and control undesirable woody vegetation and tree growth on earthen dams. Development of a basic level of tree-literacy combines basic understanding of soil properties and characteristics with basic understanding of requirements and characteristics of healthy tree root elongation and tree growth into a single conceptual understanding of management and control.

Healthy Tree Growth Requirements

The primary requirement for healthy tree growth is an environment for continual elongation of tree roots. Continual elongation of tree roots is essential to healthy tree growth for the following reasons: 1) respiration that requires a continual flow of oxygen to root tissue through soil pores; 2) soil moisture uptake that requires continual availability of soil pore water that can be captured by root tissue; 3) nutrition that requires root systems to make continuously renewed soil/root surface contact to provide needed elements and nutrients for healthy tree growth; and 4) support and stabilization that requires soil-to-root surface contact to resist externally applied loads.

Managed tree root growth control is required to prevent or minimize dangerous impacts on dams. To constrain root growth, identification of soil attributes and it's supporting environment that promote or limit growth is required. By understanding what soil conditions limit growth, various tools and techniques can be used to stop, redirect, or inhibit tree root growth and elongation. The following discussion on root growth requirements, limitations and mechanics is based on a series of publications authored and furnished by Dr. Kim Coder of the University of Georgia Cooperative Extension Services (Coder, FOR98-9, -10, -11, & -13, 1998). The reader is referred to these well-referenced publications for further and more detailed information.

Trees are not much different from all living organisms, relative to biological needs. Trees must have (1) oxygen gained through respiration, (2) water gained through adsorption and absorption, and (3) nutrition gained through adsorption and absorption, and (4) a stable foundation to withstand external forces. General root growth resource requirements are summarized in Table 1. Roots utilize soil spaces for access to water and essential element resources, and soil mass to provide structural support. Soil minerals surround the water-filled and air-filled voids or pores. These pores are continually filling and draining with water and air, depending upon the availability of water, water uptake, and atmospheric air. Root growth follows pathways of interconnected soil voids. Such voids result because of space between soil particles, between soil structural units (i.e., blocks, plates, aggregated soil, etc.); along soil fracture lines, lenses, joints, and various interstitial interfaces; and through paths of biological origins such as decayed or shrunken roots, animal burrows, etc. Better means of controlling growth can be developed by understanding resource levels that encourage and limit root growth (Coder, FOR98-9, 1998).

Root Resource	Requirements	
	Minimal	Maximum
Oxygen in soil atmosphere	2.5%	21%
Air pore space in soil (for root growth)	12%	-
Soil bulk density restricting root growth	- -	1.4 g/cc (clay) (note: 1 g/cc = 62.4 pcf) 1.8 g/cc (sand)
Penetration strength (water content dependent)	0.01 kPa (note: 1 kPa = 1kN/m^2 = 10 mbar = 0.145 psi)	3 MPa
Water content in soil	12%	21%
Root initiation (O_2% in soil atmosphere)	12%	21%
Root growth (O_2% in soil atmosphere)	5%	21%
Progressive loss of element absorption in roots (O_2% in soil atmosphere)	15%	21%
Temperature limits for root growth	40°F/4°C	94°F/34°C
PH of soil (wet test)	pH 3.5 (acidic soils)	pH 8.2 (alk. soils)

Table 1. General list of tree root growth resource requirements (After Coder, FOR98-9, 1998).

Roots survive and proliferate where adequate water is available, temperatures are warm, oxygen is present and other essential resources are concentrated. They generally tend to be shallow, limited by available oxygen and water saturation in deeper soil. However, near the base of the tree, deep-growing roots can be found, but are aerated by soil fissures and cracks and around roots where mechanical forces exerted by wind loads on the tree loosen the soil.

The ability of primary root tips to enter soil pores, open soil pores and elongate through pores is dependent upon the force generated by the root and the soil penetration resistance. As the diameter and length of an expanding root increase, its strength to resist structural failure and its expansive force it can generate both increase. The chance for structural failure increases with longer and smaller diameter roots, while short and thick roots generate significant force but minimize structural failure. Radial expansion of the root structure immediately behind the tip also helps to fracture or reduce penetration resistance in the soil ahead of the elongating root tip.

Roots use the mass of the tissues behind the tip, including root hairs, lateral root formation, and microbial entanglements to minimize the length over which root elongation force (or pressure) is expressed, thus reducing structural failure potential. As the root elongates, only root tissue within about six root diameters behind the tip is involved with force generation. Root tissue further back will act as an anchor and support base against the soil. Root tip pressure can be enormous and can range up to 9-15 MPa (9,000-15,000 mbars, 130-215 psi, or 18,700 – 31,000 psf)), with 1MPa or about 15 psi being most cited. Thus a typical root tip diameter of one millimeter is capable of generating up to about a 0.25-pound force. While tree roots cannot produce enough pressure to penetrate concrete, pipes, and most plastics or metals, they do take advantage of cracks, holes, joints and faults already in materials and exacerbate cracks and faults by growing root mass within, beneath, or around materials. When water supply is short, or when temperatures increase, diameter of roots are sacrificed to facilitate more elongation. Roots can lose more than one-third of their diameter under dry conditions, leaving roots thinner and elongating at a slower rate. Such conditions can generate passageways and set up the possibility for piping and internal erosion conditions in an earthen dam. Additionally, the loss of root contact with the soil and potential for mechanical failure of the elongating root system can lead to poor tree support, thus making a tree vulnerable to wind forces and possible upending. Tree roots are opportunistic in the colonization and control of resource space. The attributes that make a root an ideal resource gatherer for the tree conspire to make roots soil matrix explorers and fault exploiters. To prevent, control or eliminate roots from the soil infrastructure, dam owners and dam design engineers need an understanding of environmental conditions that limit and promote root growth. The foregoing discussion is summarized in terms of the four main requirements and conditions for tree growth and tree root development as follows:

Trees need to breathe. Oxygen is required for healthy tree growth through continual root elongation. In order for proper root respiration to occur, oxygen must continually move through soil pore spaces to the root tissue. Tree roots are not the only living things in the soil pore system that is competing for oxygen. As oxygen flows toward an otherwise healthy root system, enormous numbers of aerobic organisms can utilize portions, and perhaps all, of the available soil pore space oxygen before it can be utilized by root systems. If all of the oxygen is used before reaching the root system, changes must occur in the characteristics and growth rate of the root system. Trees have the ability to generate energy for short periods using carbohydrates in low or non-oxygen environments. However, this process is taxing on tree growth, and is approximately twenty times more inefficient than under normal oxygen availability and respiration conditions (Rendig & Taylor, 1989; Coder, FOR98-10,1998).   Air-filled voids in soil must be of sufficient size and continuity to allow carbon dioxide to move away from the root system and oxygen to move to the root system in order to sustain healthy root elongation and tree growth. Water-filled voids resulting from saturated soils around roots inhibit this process at a rate 10,000 times less than air-filled voids (Rendig & Taylor, 1989; Coder, FOR98-10, 1998). When oxygen drops below two to five percent of atmospheric content, root growth and the root's ability to generate elongation force significantly declines (Souty & Stepniewshi, 1988).   Table 2 summarizes air void content requirements of various soil texture and types that limit root elongation.   The table data shows that, for most embankment soils, trees need at least 10-25% air-filled voids in order to promote healthy growth.   In summary, ***Roots that cannot breath die, resulting in unhealthy, unstable, and/or dead trees.***

Soil Texture	Root-limiting % pores normally filled with air
Sand	24%
Fine sand	21
Sandy loam	19
Fine sandy loam	15
Loam	14
Silt loam	17
Clay loam	11
Clay	13

Table 2. Root growth limiting air-pore space values by soil texture (After Coder, FOR98-10, 1998)

Trees need to drink.   Second behind the need for oxygen is a tree's requirement for water.   Water uptake of trees occurs both by adsorption and absorption. In the same manner as that described for oxygen supply, tree root systems depend upon the flow of soil pore water to the root system to continually uptake sufficient water to sustain healthy root elongation and tree growth. Soil voids that are sufficiently small to prevent continual flow of pore water can limit the amount of water that elongating roots can use within the soil matrix.   Often, the moisture uptake is typically lower than that required for root elongation and healthy tree growth. As noted in Table 1, root elongation and healthy tree growth cannot be sustained where average soil moisture contents are less than about 12 percent nor greater than about 40 percent. *Soils that restrict free moisture movement preclude healthy root elongation (penetration) and healthy tree growth. Compacted soils limit pore space and therefore tend to limit supplies of both oxygen and usable water to trees.*

Trees need nourishment. Third, roots systems must provide nutrition for healthy root elongation and tree growth. Root elongation is required to encounter needed minerals, nutrients, and companion microorganisms in the soil mass. Root elongation must be continuous since replenishment of nutrients in soil is a long-term process that will not meet the requirements of stationary root systems and trees. Elongating or growing root systems continually encounter soil pores of various sizes. Soil pores that are larger than root tips create little resistance to root elongation; however, as soil pore sizes approach the size of root tips and/or become smaller than root tips resistance to root elongation increases significantly. Soil pores that are much smaller than root tips may be deformed in weak or soft soils; however, these small soil voids will reject root penetration in dense or strong soil masses. Roots cannot 'squeeze' into small, rigid soil pores within soil masses where soil strength and density preclude soil deformation and, therefore, growth is inhibited. *High strength, dense soil masses containing limited required nutrients for healthy root elongation will not sustain healthy tree growth.*

Trees need foundation support. Tree stabilization and support is provided by both components of the tree root system. The root plate (root ball) provides vertical support for the weight of the tree much the same as a shallow foundation system provides support for a building column. However, tree root systems must also resist laterally applied external loads (i.e., wind loads). Lateral root systems provide required lateral support capacity against horizontal forces through development of soil-to-root frictional forces (nature's own application of "soil nailing"). Inadequate root elongation results in reduction of base and lateral support, resulting in an unstable tree that becomes unhealthy and/or subject to failure under laterally applied loads. *Dense, compacted soil masses preclude proper lateral root elongation thus creating unstable, unhealthy trees that are subject to premature failure.*

In summary, whether in design of new dams or in maintenance of older existing dams, engineers and dam owners need to appreciate the forces, conditions and resources that control and affect the health and stability of trees so as to prevent or discourage trees from growing on new or re-constructed dams or to understand why/how trees respond to given and changing conditions on existing earthen dams.

Tree Root Elongation Management and Control

There are at least eight well-documented methods and tools available to control and limit tree (root) growth through the application of tree root elongation processes, resource availabilities, and soil preparation characteristics. These methods take advantage of depriving the tree roots of ideal resource needs for healthy growth discussed above. While these methods have been primarily used in urban or agricultural applications and settings, some methods are directly applicable to use on earthen dams and include the following methods described by Coder (FOR98-11):

1. **Intelligent designs** and applications that include techniques and materials based upon knowledge of tree growth and root development requirements. Here, minimizing available soil material faults or interfaces and tree root spaces are the preferred means for controlling and discouraging tree growth with the philosophy 'Build it correctly and they will not come!'

2. **Root kill zones** utilizing cultivation methods, sawing and cutting, trenching, vibratory plows, and chemicals to control, discourage, and remove root structure. However, these methods often result in damaging or killing the tree that, perhaps, should have been removed in the first place.

3. **Root exclusion zones** utilizing soil structure changes, soil compaction, water/aeration, stress, anaerobic conditions, soil injections and slurries, soil additives, and chemicals to prevent roots from colonizing the soil structure areas due to applied physical or chemical changes to the soil. Changing the soil structure, pore space volume or drainage/aeration matrices can generate a soil environment that roots cannot effectively grow and sustain. A variety of physical- or chemical-based soil altering materials (i.e., soil injected clay slurry or cement solutions) can be effective, at least over the short term if adequate soil volume is treated. Compacting soils appear to be a very good way to prevent root colonization. High density soils increase the resisting strength of these soils to root penetration and deprive the roots from needed oxygen and available water. Certain types of clay soils, freeze-thaw cycles, biological activity, and poor soil compaction can, over time, produce root-accessible pore space. Soil or infrastructure building material additives that neutralize or sterilize the available minerals and nutrients such as nitrogen gas, sulphur, sodium, zinc, borate, salts, or herbicides may produce serious environmental consequences, short-lived results, and non-targeted damage potential. Other methods or additives may be cost-prohibitive. See Figure 6 at end of chapter for root clearance zones.

4. **Air gap systems** designed to provide temporary and permanent air spaces for root pruning and lack of root support by use of large cobble stone barriers and drain systems. One of the more effective means of controlling tree root growth is providing stone matrices that dry quickly, create large air gaps, have poor water-holding ability, and are impermeable to systematic root penetration. Gravel layers or areas having at least 3/4 inch stone size or clean, graded, medium-sized rubble (crushed brick remnants or recycled paving and other materials), provided it is not covered or filled in with sand, are reported to produce large enough air gaps to discourage root growth.

5. **Barrier systems** using commercial root traps, root deflectors, containment devices, metals, screens, plastics, paints, and inhibitors. One of the easiest and most available materials used to control root growth are various types of 2D-type screens and barriers. While some barriers are not completely effective, many types have been shown to be effective. A list of mechanical, biological and chemical tree root growth control barriers, products and systems is shown in Table 3.

a.	Copper sulfate-soaked, synthetic, non-woven fabric
b.	Copper screen
c.	Cupric Carbonate ($CuCO_3$) in latex paint
d.	Fiberglass and plastic panels
e.	Fiber-welded geosynthetic fabric/mesh
f.	Galvanized metal screen
g.	Ground-contact preserved plywood
h.	*Geomembranes and heavy rigid plastics
i.	Infrastructure aprons and footings
j.	Metal roofing sheets
k.	Multiple layers of thin plastic sheets
l.	Nylon fabric/screen
m.	Permeable woven geosynthetics
n.	Rock-impregnated tar paper/felt
o.	*Slow-release chemical barriers
p.	Thin layered bitumen & herbicide mixtures
q.	Woven and non-woven slit-film plastic sheets

*Common commercial tree growth control products available

Table 3. Selected list of tree root growth control barriers (after Coder, FOR98-11, 1998)

The costs of these products will likely continue to decrease as the demand for these products increases in the future. Of the barriers shown in the list, three types are most commonly used: traps (root engaging and constricting), deflectors (walls), and inhibitors (chemical constraints). Combined features of the barrier, the site, and barrier installation and maintenance are critical to their effectiveness, but no barrier should be assumed to stop all roots under all conditions. Most types of mechanical and chemical barriers have limited effectiveness lives and this should be factored into any long-term cost analysis. The reader is directed to the Table 3 reference source and other related publications for details on commercially-available root barriers.

6. **Directed growth systems** to concentrate roots in desired directions, guide root growth along channels, allow root survival in desired areas, and create root culverts or layers. As noted earlier, roots are opportunistic and grow and proliferate where there are good supplies of resources. Understanding root elongation, colonization, and survival processes allows growth-favoring soil layers, corridors, and areas to be designed for directing or deflecting roots away from infrastructures where tree roots can be harmful. Several methods or systems are used to attract, deflect, channel or lead roots in a direction or area as needed. One attraction method used is called "baiting" and involves providing ideal essential soil condition resources in a direction away from an infrastructure. The net result is a much higher survival and growth rate in that part of the root system as opposed near infrastructures where root damage can occur. Water, growth nourishment elements, and oxygen should be limited and compaction should be maximum near infrastructures.

    Another method is to "shepherd" roots to desirable locations using trenches, channels, layers, raceways, tunnels, and other devices that are surrounded by root control obstacles, barriers, or resource constraints. Growth channels filled with rich, well-aerated, ecologically healthy growth medium will encourage root colonization and survival in areas away from sensitive infrastructure targets.

7. **Selection of desired species** of trees that require lower soil oxygen environments, have improved root morphology, and are more effective species for long-term solutions. This method focuses on choosing and planting available tree species that can survive under rather limited or harsh environmental conditions. Several tree species are available that are small in size, have shallow and less aggressive rooting, and are slower growing. Dam owners, however, should be reminded again that trees in general are not a good plant option and have no place on dams; instead, more desirable, native grasses should be planted and maintained.

8. **Creating avoidance zones** to separate tree growth from earthen dam embankments and dam appurtenances where root damage may be critical thus establishing biological-free zones that reduce potential problems. This method simply recognizes that there are places where trees are acceptable and other places, namely dams, where they are not (see Figure 6).

The most practicable of these methods for use on earthen dams are those associated with intelligent design development, exclusion zones, kill zones, and barriers. Within this group of suitable methods the combination of intelligent design development and exclusion zones are the most effective. With an understanding of the previous meshing of soil properties with healthy tree root elongation, it is not difficult to develop an intelligent design scheme for new dams and the remedial repair of existing dams. An intelligent design philosophy associated with dam embankment design and construction would involve proper embankment soil compaction as the means of exclusion of root elongation.

In summary, there are many tools, methods and options for minimizing or preventing tree root-caused damage to earthen dams. The most important management (and design) concept to understand is how tree roots are invited to be associated with interstitial elements and colonize soil matrices and discontinuities, and resource availability areas. Our responsibilities as owners and dam design engineers must lie with creating and using any or a combination of the numerous root growth control tools and techniques that are tree-literate so that trees do not have the opportunity to become a safety problem to embankment dams and their appurtenances in the first place.

Exclusion by Embankment Compaction

Design and construction practices of using optimum compaction of embankment soils reduce potential settlement of embankments, increases shear strength of the embankment soils, decreases the permeability of the embankment soils, and minimizes long-term changes in the physical and engineering properties of soils. When embankment soil compaction results in the attainment of desirable objectives from a geotechnical engineering behavior perspective of earthen slopes, compaction of embankment soils also precludes tree root growth and elongation as a result of exclusion of most of the requirements for healthy root elongation and tree growth. As has been previously noted, densely compacted soils discourage root elongation through increased resistance, lowered oxygen levels, and reduced available water. Traditional embankment soil compaction specifications require that the soil be compacted to about 95 to 98 percent of the standard Proctor maximum dry density as determined by ASTM D-698. Furthermore, most properly written soil compaction specifications generally require that compaction moisture contents be maintained about two percent below to three percent above optimum moisture content. At these degrees of compaction and at these moisture contents, soil oxygen content, water content, and soil pore size are not available for healthy root elongation and tree growth. Even if there is sufficient moisture content in the soil to otherwise sustain healthy root elongation, the soil pore sizes are so small that available pore water cannot be effectively moved to the root system. Consequently, the compacted dam embankment fill soil produces an exclusion system that mechanically impedes healthy root elongation and tree growth. Table 2 provides a summary of minimum air voids for various soil types required to impede root elongation for healthy root and tree growth.

Maintenance Mowing and Kill Zones

The second most effective method of controlling woody vegetation and tree growth on dam embankments is through the use of native grass or ground cover with maintenance mowing, and using kill zones where necessary around critical structures to control trees and other undesirable nuisance-types of vegetative growth. Maintenance mowing should be done *at least* twice per year with one mowing scheduled for spring after initiation of new spring growth and the second mowing scheduled for late fall immediately prior to the first killing frost or freeze (See Chapter 7). The spring mowing should be a very close cutting of all vegetation to allow maximum sunlight to penetrate to desirable grass cover species. The fall cutting should not be as close as the spring cutting to provide maximum resistance to surface runoff erosion and to provide cover for desired wildlife species (quail, rabbit, grouse, songbirds, etc.).

In areas where regular maintenance mowing is not practical to control woody vegetation and tree growth, the selective use of herbicides might become necessary to control small woody vegetation and tree growth. There are many commercially available herbicides that are environmentally safe to use in most applications. However, one must always be careful in the use, or overuse of herbicides, because they are design to kill and/or impede (slow) plant growth. Overuse of herbicides may contaminate areas of the dam embankment to such an extent that desirable grass cover cannot be effectively grown. One must always follow manufacturers recommendations when using herbicides, or better yet, solicit the advice of the nearest USDA/NRCS agent prior to using herbicides to control woody vegetation and tree growth on earthen dams.

Chemical Barrier Systems to Inhibit Root Growth

Commercially available barrier systems are effective in controlling root elongation and growth; however, many of these barrier systems are relatively expensive and cannot be justified for placement over the entire earthen dam embankment. These barrier systems are often economical for placement on portions of earthen dams where accessibility is difficult after construction and/or where particularly problematic and nuisance woody vegetation and tree growth is likely to occur.

One typical biocide product, called "Biobarrier©" is marketed and promoted, among other applications such as sidewalk and landfill cap protection, to prevent tree and plant roots from penetrating dams. The product consists of long-term, slow release nodules containing Trifluralin herbicide, that are bonded to a geotextile fabric as shown in Figure1.

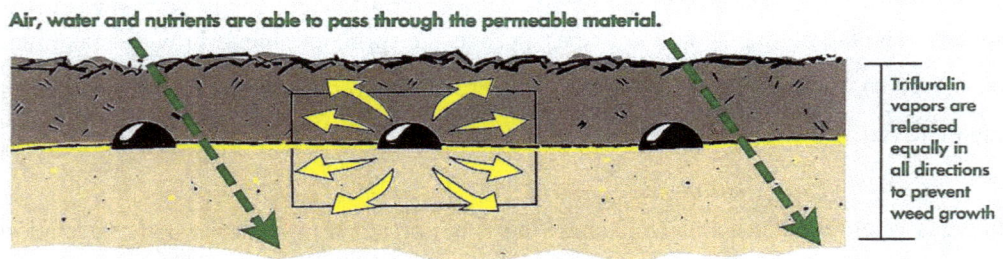

Figure 1. Chemical biocide barrier installation showing slow-releasing biocide nodules attached in a woven fabric matrix and installed under a cover of soil, mulch, gravel or stone (Biobarrier©).

This particular barrier is environmentally acceptable to EPA and indicated to be effective against all types of roots around pipes, hardscapes, and dams and levees. While the product is guaranteed for 15 years, its life is inversely proportional to environmental temperature conditions.

For example, its effectiveness is expected to be about 40 years at 20°C (68°F) and about 100 years at 15°C (60°F). For deep soil cover, it is expected to last 100 years; for near soil-surface weed control installations, where temperatures are higher and cycle daily, the projected life is expected to exceed the guaranteed 15-year life. Figures 2 and 3 show an application of this product on a 25-foot high and 350-foot long earthen dam to prevent deep penetration of deep-rooting native trees and woody vegetation such as willows, sagebrush, and chokecherries.

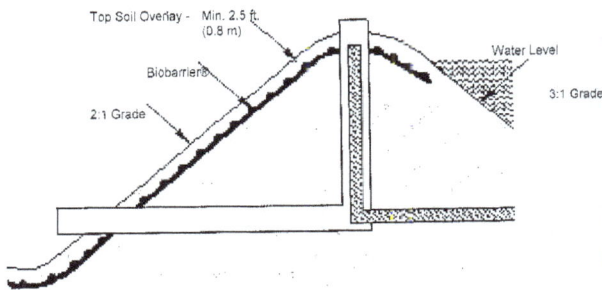

Figure 2. Earth dam installation of chemical

Figure 3. Installed chemical barrier on a dam in Montana (Kershner, 1992)

Herbicidal Applications

Herbicidal delivery to control undesirable vegetation depends on several considerations which include (a) types of plants and weeds (herbaceous, vines, trees, brush, phreatophytes, etc.), (b) site conditions (geology/sinkholes/karst), topography, (c) proximity to water bodies, (d) riparian land use, (e) sensitive environmental factors (Federal, state & local regulations; potential off-site wind drift over water or land), and (f) application factors (dosage, placement, retention time, plant growth stage, physiological factors, and method of application). A very important consideration is for the user to follow the herbicide manufacturer's warnings and instructions. The user is also encouraged to consult with a local county extension office or agent to obtain advice on the best and safest herbicide to use and on what recommended application technique to use. While there are several herbicidal delivery methods available, the most common techniques are shown below in Figure 4.

- Foliage spraying
- Tree injection
- Frill or girdle treatment (slash through bark then spray or paint)
- Basal bark spraying
- Cutting tree and poisoning stump
- Soil treatment
- Other

Figure 4. Herbicide delivery application methods

Some of these techniques and herbicides used are illustrated in Figures 5a – 5f. The U. S. Department of Agriculture (SCS, now NRCS) published a useful methods, treatment points, and time of treatment guidelines for controlling trees and brush on dams, including some of the applications listed in Figure 4 (USDA, 1988). Table 4 summarizes the USDA recommendations. With the exception of Krenite, which is applied to the foliage, 2,4-D is the only approved herbicide for poisoning trees on dams. 2, 4-D is manufactured by several companies and is sold under several trade names. In all cases, the user is cautioned again to follow the manufacturer's instructions and should consider the manufacturer's label instructions to supercede recommended instructions in the USDA table.

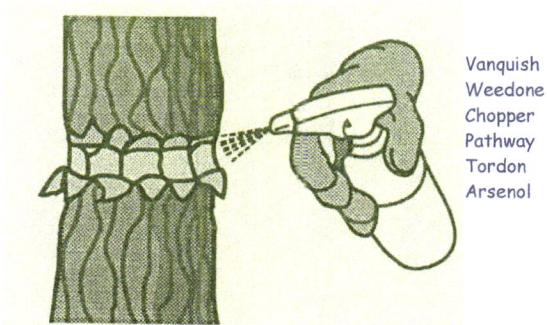

Vanquish
Weedone
Chopper
Pathway
Tordon
Arsenol

Figure 5a.   Frill cut application.

Touchdown
Crossbow
Access

Lower 18":
Ester derivatives
to penetrate bark

Figure 5b.   Basal bark spraying.

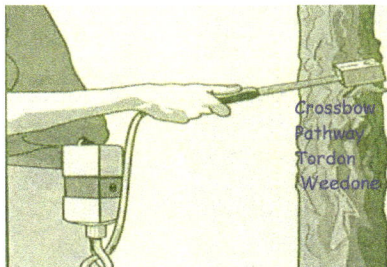

Crossbow
Pathway
Tordon
Weedone

Figure 5c.   Hypohachet application.

Apply treatment
within an hour
of cutting tree

Figure 5d. Cut-stump application.

Vanquish
Weedone
Weedmaster
Crossbow
Banvel
Roundup
Accord
Arsenol
Chopper
Pathway
Tordon
Touchdown
Spike
Garland

Figure 5e.   Backpack foliar application on dam.

Figure 5. Applications and techniques for different herbicidal deliveries to trees and brush, with
example commercially-available herbicides listed.

Figure 5f. Tractor spraying application on dam.

Method of Application	Recommended Time
• Cutting trees and poisoning stumps • Injection • Foliage spraying • Frill treatment (trees larger than 4" dbh) • Basal spraying (trees smaller than 6" dbh) • Prescribed burning (trees smaller than 2" dbh)	• Growing season • Anytime • Last two (2) months of growing season • Anytime • Growing season • See technical specifications

Table 4. Recommended methods and time of herbicide treatment application (USDA, 1988).

USDA recommends that trees killed by herbicide should be removed within the year following treatment to prevent front slope from falling into the reservoir and plugging the spillway. Downed trees on the back slope should also be removed to prevent potential problems of seepage, erosion, burrowing animals, etc.

The reader is referred to the USDA guideline for detailed discussion on each of the six treatment methods listed in the above table. These methods can be applied to establish tree and woody plant clearance or avoidance zones on and around dams as illustrated in Figure 6.

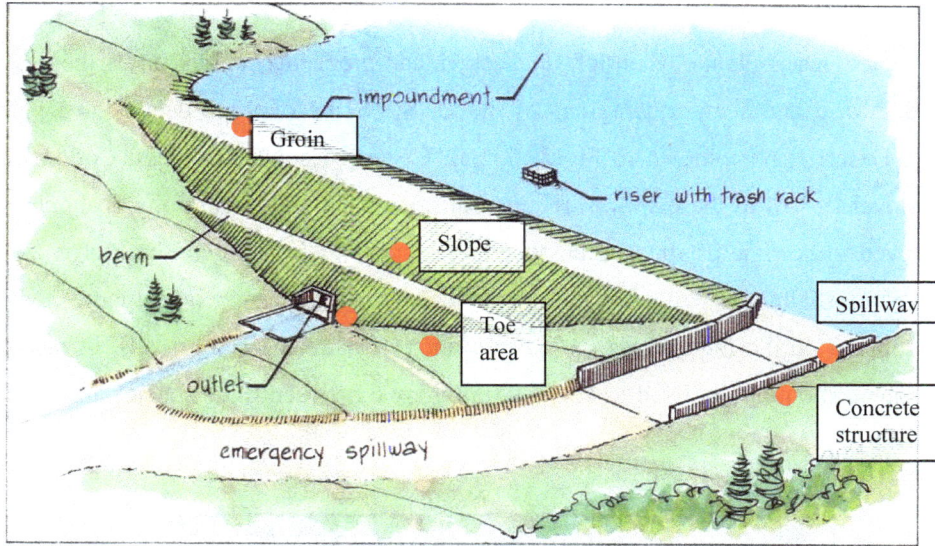

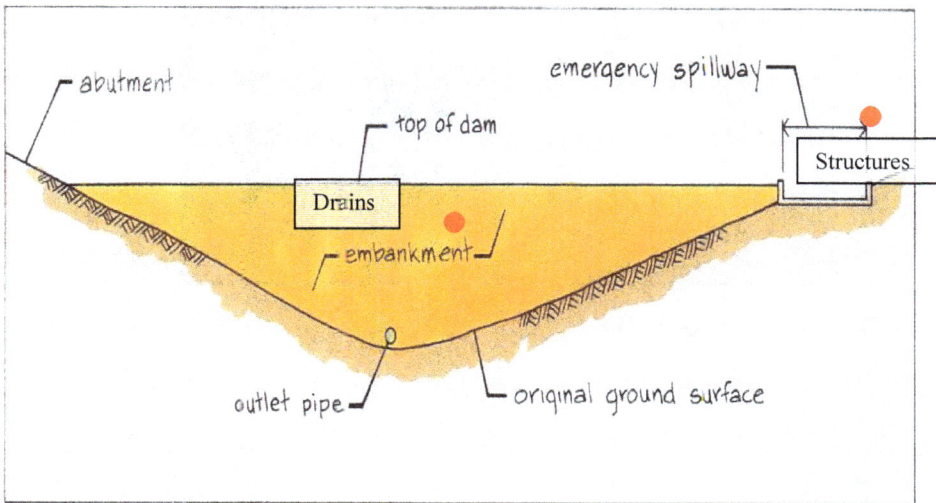

Figure 6. Tree clearance zones for embankment dams and dikes.

Desert Plants

Deep-rooted desert plants, when left unchecked, can propagate rapidly on earthen dams located in arid and semi-arid regions of the U. S. Some of these deep-rooted plants include Desert Broom shown in Figure 7, Salt Cedar, Mesquite, Cypress, Cottonwood and Paloverde. All of these species require considerable effort to control and should not be allowed to become established anywhere on dams. Palm trees can be a problem in that they are shallow-rooted, but develop a large root ball that can produce large cavities when toppled during high winds. Upstream and downstream access roads, in place at many dams, should be utilized to create a buffer zone between these species and the toe of dams.

Figure 7. Deep-rooting Desert Broom Plant

The Maricopa County, Arizona, Flood Control District (MCFCD) recommends, in cases where deep-rooted plants are two feet in height or less, that they be controlled with a 3-5% solution of Roundup® Pro (Renckly and Drake, 1999). If the plants are over two feet in height they should be hand cut to ground level. The stumps should be treated within the first five minutes by an almost straight mix of either Roundup Pro® or Garlon 3A-Garlon 4®, depending on the temperature conditions. MCFCD recommends that when treating Salt Cedar near waterways that Rodeo be sprayed at a 3-5% solution with six ounces of Siltwet® per acre added. This is sprayed on plants two feet in height and under. Plants over two feet are hand cut and the stump treated with an almost straight solution of Rodeo® within five minutes of cutting the plant.

Revegetation on earthen dams is recommended to minimize erosion on the embankment slopes and to provide natural landscaping for earthen dams. MCFCD recommends hydro-seeding over labor-intensive hand-seeding to revegetate dam embankments. Figure 8 illustrates hydro-seeding operations on a floodway dam. Seed, water, tack material and a wood fiber or paper mulch are mixed in a hydro-seeder and sprayed directed onto the slopes. The seeds are encapsulated in the mulch and tack material until enough moisture is present to begin the germination process.

Figure 8. Hydro-seeding operations on a floodway dam in Maricopa County, Arizona (Renckly & Drake, 1999).

MCFCD has found that it takes 2 to 3 years before "significant" vegetative cover results are achieved because of the arid climate and high degree of embankment compaction. MCFCD determines the desirable seed mix by first laying out a test acre on the dam embankment and a plant count is then taken of all the different plant species that are native to the area and placed on the test acre. This plant count is converted by the seed supplier into the amount of seed needed to germinate the desired amount of the species per acre. The amount of pure live seed (PLS) applied for individual plant species also varies by availability from the local seed supplier. Table 4 shows a seeding mixture specified for one of the District's dams and is typical of specified hydro-seeding mixes. No deep-rooted species are allowed in the seed mix.

MCFCD has found that revegetation efforts have successfully reduced erosion problems, but has attracted both desirable and undesirable animals.

SEEDING MIXTURE

Common name	Scientific Name	Pounds of Seed Per Acre
Purple three-awn	Aristida purpurea	4
Indian Wheat	Plantago insularis	3
Needle Grama	Bouteloua arstiodoides	1
Desert Marigold	Baileya multiradiata	1
Mexican Gold Poppy	Eschschotzia mexicana	1
Creosote	Larrea tridentata	8
Brittle Bush	Encelia farinosa	2.5
Bursage	Ambrosia deltoidea	2

** **Note:** Apply 1500 pounds of wood fiber mulch in Hydro-seed mix, plus 150 – 200 pounds tack material per acre.

Table 4. Typical seed list specified for a flood control dam managed by the Maricopa County, Arizona, Flood Control District (Renckly & Drake, 1999)

References:

1. Coder, K. D., Tree Root Growth Control Series: Root Growth Requirements and Limitations, Univ. of Georgia, Cooperative Extension Service Forest Resources, Publication FOR98-9, March 1998, 8 pp.

2. Coder, K. D., Tree Root Growth Control Series: Soil Constraints on Root Growth, Univ. of Georgia, Cooperative Extension Service Forest Resources, Publication FOR98-10, March 1998, 8 pp.

3. Coder, K. D., Tree Root Growth Control Series: Methods for Root Control, Univ. of Georgia, Cooperative Extension Service Forest Resources, Publication FOR98-11, March 1998, 9m pp.

4. Coder, K. D., Selected Literature: Root Control Methods, Univ. of Georgia, Cooperative Extension Service Forest Resources, Publication FOR98-13, March 1998, 4 pp.

5. Coder, K. D., Root Growth Control: Managing Perceptions and Realities, Proceedings, Second International Workshop on Tree Root Development in Urban Soils, International Society of Arboriculture, March 5-6, 1998, edited by D. Neely and G. Watson, pp. 51-81.

6. Coder, K. D., Engineered to Fail? Tree Root Management on Dams, Abstract, University of Georgia, Athens, November 1999, 1 page.

7. Rendig, V. V. and H. M. Taylor, Principles of Soil-Plant Interrelationships, McGraw-Hill, 1989.

8. USDA-SCS, Technical Notes (OK-8), Control of Trees and Brush on Dams, Stillwater, Oklahoma, April 5, 1988.

9. Sisneros, D., USDI-USBR, Res. And Lab. Serv. Div., Upper Colorado Region Saltcedar Cost Analysis, Memo 94-2-2, February 1994.

10. USDI-USBR, Water Operation and Maintenance, Bulletin No. 150, Guidelines for Removal of Trees and Vegetative Growth from Earth Dams, December 1989.

11. Biobarrier©, Application Manual, Root Control System, BBA Nonwovens/Remay, Inc. Product Information, August 1999, Old Hickory, Tennessee.

12. New Hampshire DES Environmental Fact Sheet, <u>Tree Growth on Dams</u>, WD-DB-8, 1997.

13. Ohio Department of Natural Resources, <u>Dam Safety: Trees and Brush</u>, Fact Sheet 94-28, July 1999.

14. Pennsylvania DEP, <u>Fact Sheet – Vegetation/Erosion Control on Dams</u>, 31-40 FS,DEP1909,June1997,http://www.dep.state.pa.us/dep/deputate/watermgt/WE/FACTS/fs1909.htm.

15. Renckly, T., Drake, G., <u>Plant & Animal Management Practices on Flood Control Dams</u>, Maricopa Co., Arizona, November 1999.

16. USDA-SCS (NCRS), S. Tech. Serv. Ctr., Technical Note 705, <u>Operations & Maintenance Alternatives for Removing Trees from Dams</u>, April 1, 1981.

17. Univ. of Tenn. Agr. Extn. Serv., R. Bullock, <u>Chemical Vegetation Management on Non-cropland</u>, Bulletin PB-1538, December 1995.

18. Univ. of Tenn. Agr. Extn . Serv., G. Rhodes & G. Breeden, <u>2001 Weed Control Manual for Tennessee</u>, Bulletin PB-1580, December 2000.

19. Kershner, C., <u>Geotextiles - It's Only Natural</u>, Land and Water, January 1992.

20. STS Consultants Ltd., ASDSO Working Group, <u>Dam Safety Guidebook</u>, 1985.

21. Association of State Dam Safety Officials (ASDSO), Report on Specialty Workshop #1: Plant & Animal Impacts on Earthen Dams, Knoxville, Tennessee, November 30-December 2, 1999, June 2000.

Chapter 6
Dam Remediation Design Considerations

Specific dam remediation design considerations, procedures, and techniques will be considered for each of the previously identified dam safety inspection and evaluation zones. Figure 1 presents these zones as a review prior to discussion of potential dam remediation design considerations for each zone. Dam remediation design alternatives presented herein should be considered examples. These remediation design examples should not be considered the only alternatives for use in dam remediation design to correct deficiencies associated with tree and woody vegetation growth on earthen dams. Some additional dam remediation design alternatives presented for correction of tree and woody vegetation growth related deficiencies also provide positive correction of other deficiencies and protection against other types of earthen dam deterioration.

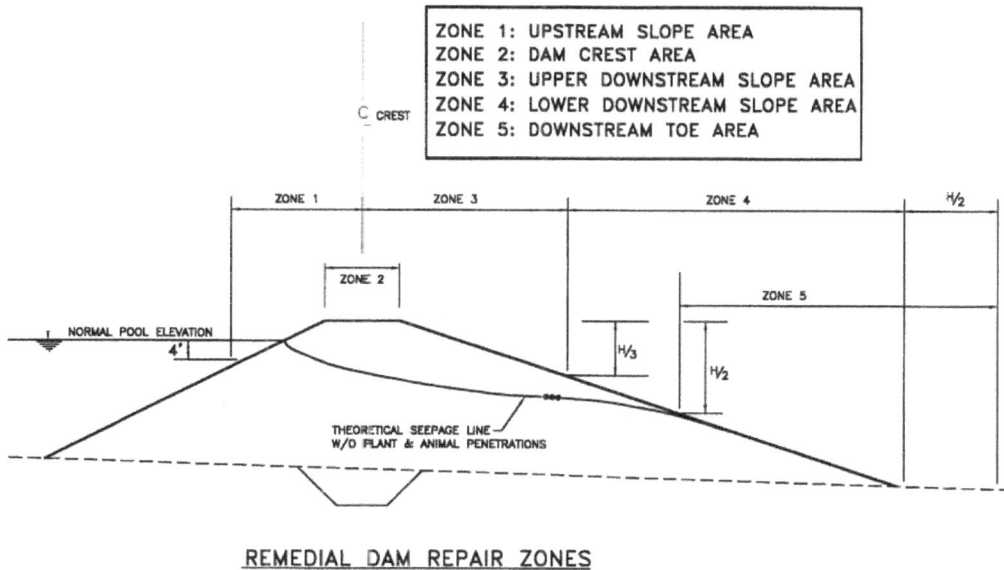

```
ZONE 1: UPSTREAM SLOPE AREA
ZONE 2: DAM CREST AREA
ZONE 3: UPPER DOWNSTREAM SLOPE AREA
ZONE 4: LOWER DOWNSTREAM SLOPE AREA
ZONE 5: DOWNSTREAM TOE AREA
```

REMEDIAL DAM REPAIR ZONES

Figure 1

Inspection and Evaluation Zone 1

Figure 2 illustrates potential problems that can occur in Zone 1 with respect to tree and woody vegetation growth on earthen dams. This illustration also depicts the occurrence of wave erosion, vehicle access, and surface runoff erosion. Potential problems illustrated include instability of relatively large trees on the upstream slope and dam crest, and alteration of the seepage line as a result of wave erosion.

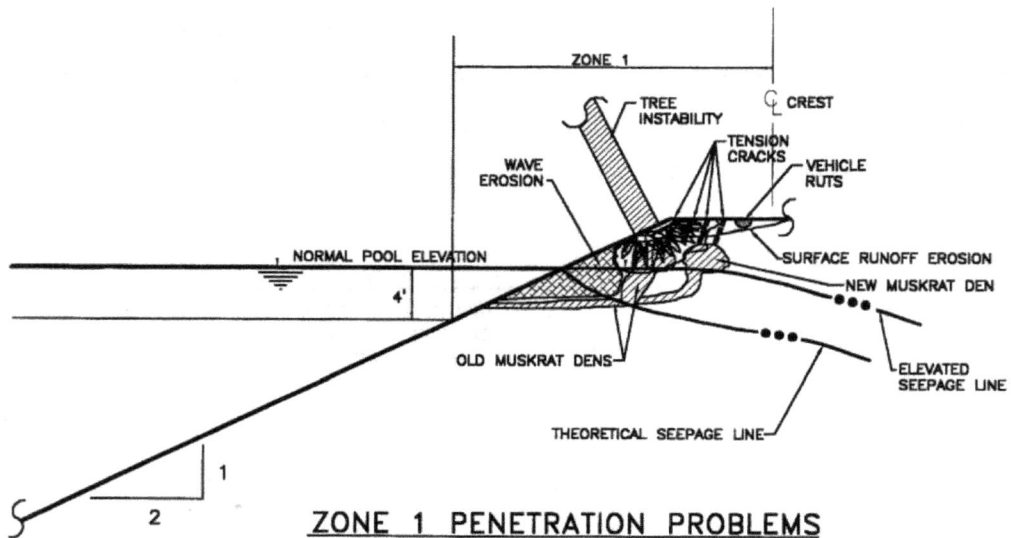

ZONE 1 PENETRATION PROBLEMS

Figure 2

Dam remediation design techniques necessary to address potential problems illustrated in Figure 2 are illustrated in Figures 3 and 4. Dam remediation construction typically requires lowering of the normal pool elevation and/or complete drawdown of the retained reservoir. This is particularly true for dam remediation construction in Zone 1. The normal pool elevation should be lowered as far ahead of the scheduled dam remediation construction as practicable.

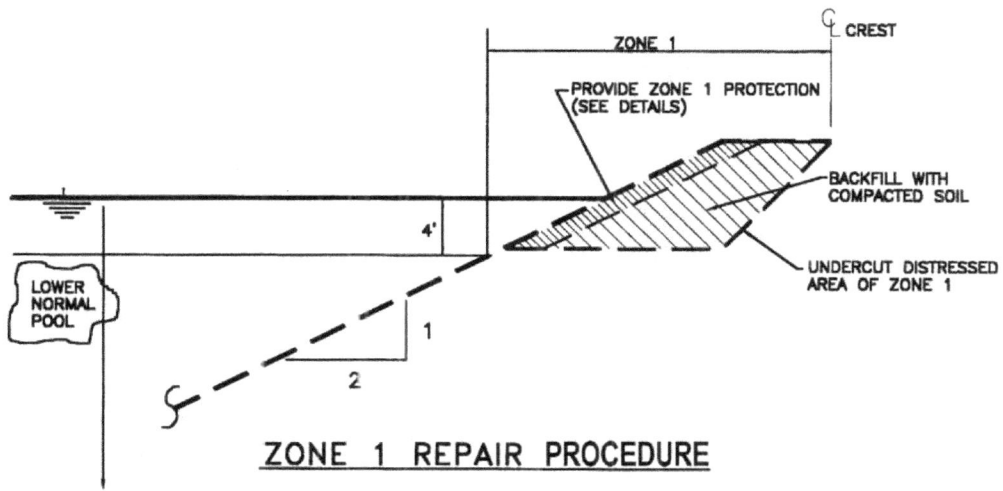

ZONE 1 REPAIR PROCEDURE

Figure 3

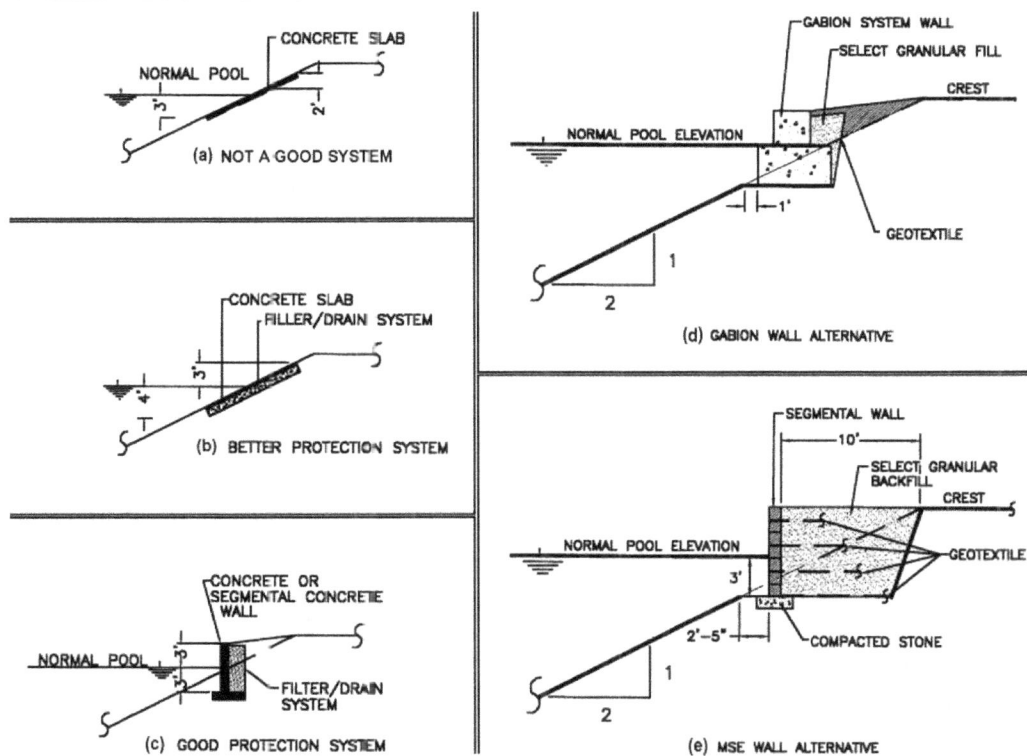

REMEDIAL REPAIR DESIGN
ALTERNATIVES FOR ZONE 1

Figure 4

Tree and woody vegetation growth in Zone 1 must be undercut to remove all stumps, rootballs, and root systems developed by tree penetrations as illustrated in Figure 3. The required depth of undercutting typically extends to near the limits of Zone 1, which is about four feet below normal pool elevation. In the case of earthen dams with narrow crest widths, the backslope of the undercut area will typically extend to near the centerline of the dam crest or the downstream limits of Zone 1. Subsequent to undercutting affected areas of Zone 1, the undercut area must be thoroughly inspected to confirm that all major root systems (greater than about one-half inch in diameter) have been removed during the undercutting operation. Following inspection and approval of the undercut area by the engineer, suitable backfill should be placed in the excavation and properly compacted to the dam remediation design limits. Backfill should consist of approved embankment fill material and should be compacted to a minimum of 95 percent of the maximum dry density of the fill soil as determined by the standard Proctor compaction test (ASTM D-698). In conjunction with the undercutting and backfilling, the dam remediation design should include a slope protection system to deter future tree and woody vegetation growth and reduce the potential for wave and surface runoff erosion.

Figures 4(a) through 4(c) illustrate various configurations of rigid (concrete) upstream embankment slope protection systems. Figure 4(a) illustrates a concrete slab being placed directly on the upstream slope from about three feet below to about two feet above normal pool elevation. While this system is somewhat limited relative to the area of protection, the most critical aspect of this system is that it provides no filtration and/or drainage system beneath the concrete slab. Continual wave action and the buildup of hydrostatic pressures beneath the concrete slab will eventually result in downward movement of the slab. Figure 4(b) illustrates a better dam remediation design utilizing a concrete slab slope protection system. This slope protection system has been improved over the original system by covering a larger area of the upstream slope and by providing a filter system beneath the concrete slab protection system. The author is of the opinion that the dam remediation protection system shown in Figure 4(c) is the most desirable and cost effective design for use of reinforced concrete for a protection system. The reinforced concrete wall provides a gentle slope to flat backfill area that can easily be maintained by mowing to preclude tree and woody vegetation growth. In addition, this dam remediation design alternative can be used to provide a wider effective dam crest and provides excellent protection against wave erosion.

NOTE: Reinforced concrete wall and slab systems constructed on the upstream slope must always be provided with filtration/drainage systems to reduce the potential for development of excessive hydrostatic pressures and internal erosion and scour of soil from beneath the structures. The referenced figures are presented for illustrative purposes and should not be used for actual dam remediation design without proper design analyses to confirm any indicated dimensions of the drawings.

Alternative flexible upstream slope protection system designs for use in Zone 1 are shown in Figures 4(d) and 4(e). The author has utilized both of these flexible slope protection systems effectively to reduce potential tree and woody vegetation growth on upstream slopes and to provide resistance to wave and surface erosion. Figure 4(d) illustrates a typical gabion wall system while Figure 4(e) illustrates the use of a Mechanically Stabilized Earth (MSE) wall system for protection of the upstream slope of an earthen dam.

NOTE: Granular backfill material used in design and construction of these flexible wall systems must be protected against soil contamination and internal erosion of retained soil by an effective geotextile filter/drainage material and/or a graded aggregate filter. These figures are presented herein for illustrative purposes and should not be used for actual design without proper design analyses to confirm any indicated dimensions of the drawings.

Inspection and Evaluation Zones 2 and 3

Potential problems associated with tree and woody vegetation growth on earthen dams in identified Zones 2 and 3 are illustrated with dam remediation design procedures in Figure 5. Potential problems illustrated for Zone 2 include the growth of mature trees having stump diameters greater than twelve inches. Mature trees having stump diameters greater than eight inches are illustrated at various locations throughout Zone 3 and in the overlap area of Zones 2 and 3.

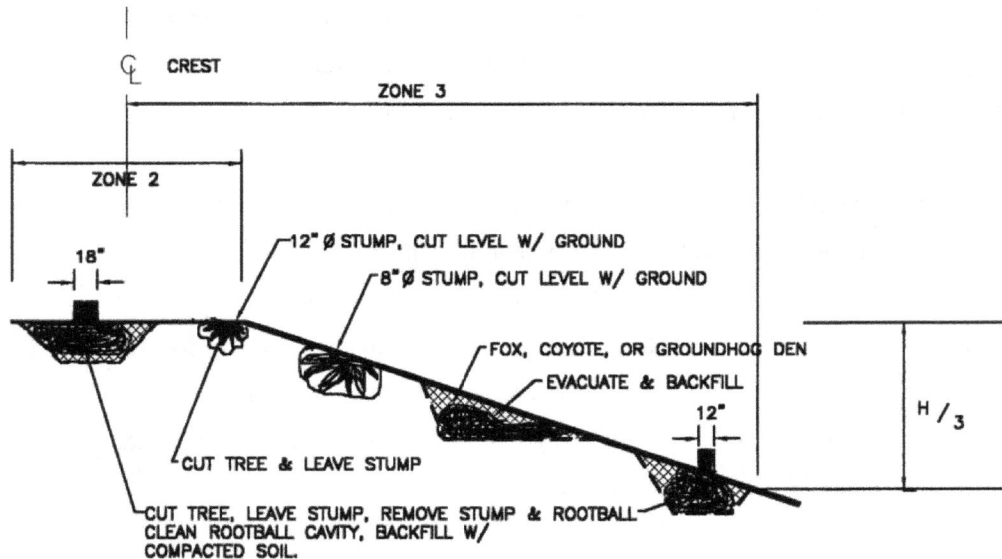

ZONE 2 & 3 REPAIR PROCEDURES

Figure 5

Two dam remediation design procedures are illustrated in Figure 5 for removal of trees of various sizes. This illustration implies that trees located in the overlap area of Zones 2 and 3 having stump diameters less than about twelve inches could be cut flush with the ground and left in place for future treatment of the decayed stump and rootball system. However, removal of all stumps, rootballs and root systems is always the better and more conservative approach to removal of mature trees. Subsequent to cutting of trees having stump diameter less than about twelve inches in the overlap area of Zones 2 and 3, the surface of the stump can be treated with a protective coating similar to polyurethane that will prolong the decaying process. Conversely, the referenced illustration indicates that any trees in Zone 2 upstream of the overlap area of Zones 2 and 3 having stump diameters of twelve inches or greater should be treated by total removal of the tree, stump, rootball, and root system. The suggested dam remediation design and construction procedure suggested for complete removal of trees, stumps, rootballs, and root systems in Zones 2 and 3 consists of the following activities:

1.   **Cut** the tree approximately two feet above ground leaving a well-defined stump that can be used in the rootball removal process;

2.   **Remove** the stump and rootball by pulling the stump, or by using a track-mounted backhoe to first loosen the rootball by pulling on the stump and then extracting the stump and rootball all together (this is much the same procedure a dentist would use in extracting a tooth);

3.   **Remove** the remaining root system and loose soil from the rootball cavity by excavating the sides of the cavity to slopes no steeper than 1:1 (horizontal to vertical) and the bottom of the cavity approximately horizontal; and

4.   **Backfill** the excavation with well-compacted soil placed in relatively thin lifts not greater than about eight inches in loose lift thickness. Compaction of backfilled soils in these tree stump and rootball excavations typically requires the use of manually operated compaction equipment or compaction equipment attached to a backhoe.

NOTE: All disturbed areas must be protected by seeding and mulching.

Figure 5 further illustrates that trees located in Zone 3 that have stump diameters greater than about eight inches should be treated by total removal. The removal procedure should be the same as previously described for larger trees in Zone 2. Trees having stump diameter of less than about eight inches could be cut flush with the ground and treated with a waterproofing sealant similar to polyurethane to prolong the stump and rootball decaying process. Again, complete removal of the stumps, rootballs, and root systems of all mature trees is a better and more conservative method of remediation.

Inspection and Evaluation Zone 4

Figure 6 illustrates potential problems associated with tree and woody vegetation growth in Zone 4 of an earthen dam with suggested dam remediation design and construction procedures.

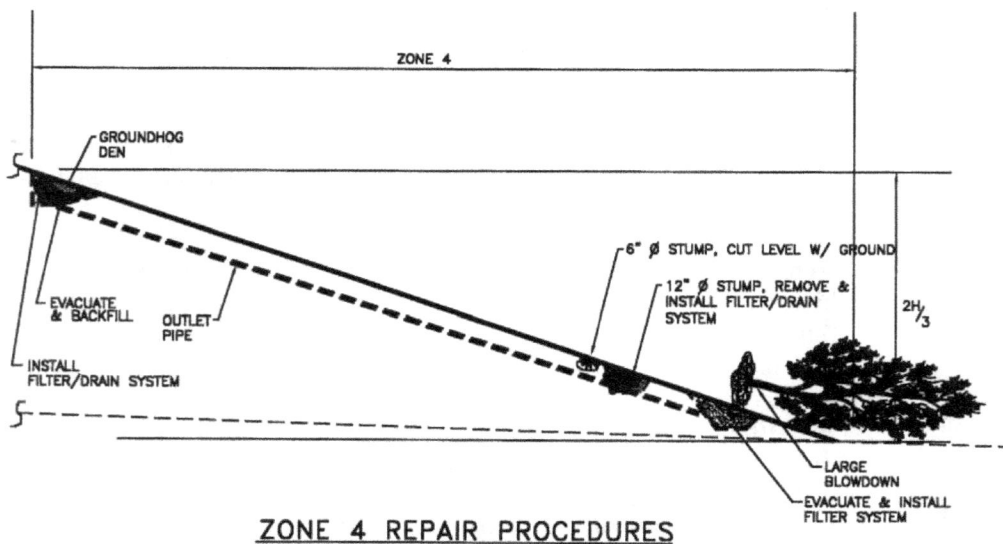

ZONE 4 REPAIR PROCEDURES

Figure 6

Young immature trees having stump diameters less than about six inches can be removed by cutting flush with the ground and treating the stump with a wood preservative and/or sealant to prolong the decaying process. This procedure is based upon the fact that immature trees of this size typically have not developed a rootball and/or root system that will significantly impact the zone of saturation or the seepage line in Zone 4.

Trees having stump diameters greater than about six inches must be treated by complete removal; however, the dam remediation design and construction procedure for total removal of trees in Zone 4 is somewhat more complicated than total removal of trees in previously discussed zones. Treatment of mature tree penetrations in Zone 4 involves the following activities:

1. **Cut** the tree approximately two feet above ground level leaving a prominent stump for use in the rootball extraction process;

2. **Remove** the stump and rootball by pulling the stump or extracting with a track-mounted backhoe after loosening the rootball by pulling on the stump from different directions;

3. **Clean** the rootball cavity to remove loose soil and the remaining root system by excavating the rootball cavity with maximum 1:1 (horizontal to vertical) side slopes and a horizontal bottom; and

4. **Install** a subdrain and/or filter system in the tree penetration excavation and backfill with compacted soil placed in maximum loose lifts of eight inches.

Note: Backfill placed in all tree removal excavations must be compacted to a minimum of 95 percent of the maximum dry density as determined by ASTM D-698.

Note: Subdrain and/or filter systems installed in tree removal excavations in Zone 4 may be incorporated into major subdrain systems to be installed in the overlap area of Zones 4 and 5.

Inspection and Evaluation Zone 5

The author identified Zone 5 as one of the two most critical zones for tree and woody vegetation growth on an earthen dam. Figure 7 illustrates some of the problems that can occur with tree and woody vegetation growth in Zone 5. The major adverse feature in Zone 5 is typically the interception of the downstream embankment slope by the seepage line. The author is a strong advocate of the installation of embankment subdrain systems during dam remediation design and construction even though the earthen dam may have been provided with an embankment subdrain system during original design and construction.

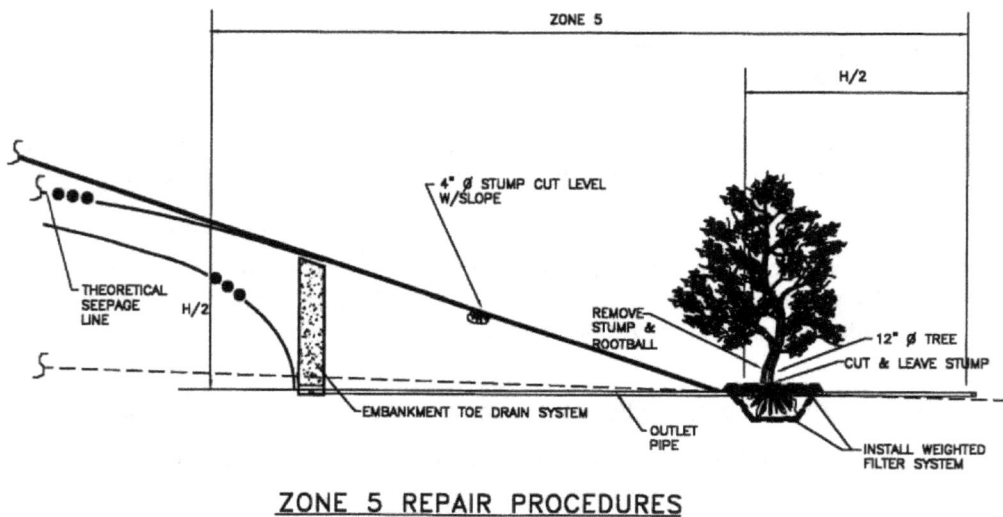

ZONE 5 REPAIR PROCEDURES

Figure 7

One must understand the impact of tree removal in Zone 5 on the seepage line and the quantity of seepage that will occur subsequent to dam remediation in this zone. As indicated by Figure 7, trees in Zone 5 having stump diameters less than about four inches can be cut flush with the ground and the stump treated with a waterproof sealant to prolong stump and rootball decay. Trees having stump diameters greater than about four inches must be removed completely. If the embankment toe drain or subdrain system is installed in advance of tree removal in Zone 5, the rootball cavity can be backfilled with compacted soil, provided seepage does not emerge from the excavation and/or the tree is located beyond the toe of the embankment slope. Tree rootball cavities existing beyond the toe of the downstream embankment slope generally require the installation of a filter system and in some cases a weighted filter system as indicated in Figure 7. The weighted filter system may be converted to a weighted drain system by installing a drain and outlet pipes connected to the outlet pipe of the embankment subdrain system.

Summary of Dam Remediation Design Considerations

A summary of dam remediation design considerations for treatment of tree and woody vegetation on earthen dams is presented below. Dam remediation design procedures and techniques are presented for treatment of various size trees in the identified dam safety inspection and evaluation zones.

Remedial Repair Zone	Procedures and Techniques
Zone 1	Remove all trees, stumps, rootballs, and root system; clean rootball cavity; and backfill with properly placed and compacted soil backfill. Install tree and woody vegetation and wave erosion protection system on the upstream slope from about four feet below normal pool elevation to about three feet above normal pool elevation.
Zone 2	Cut trees in overlap area of Zone 2 and Zone 3 having stump diameters of twelve inches or less flush with the ground and treat the stump with a waterproof sealant to prolong stump decay.

Completely remove trees having stump diameters of about twelve inches and greater, and backfill rootball cavity with properly compacted backfill soil.

Zone 3

Cut trees having stump diameters of about eight inches and less level with the ground and treat the stump with a waterproof sealant to prolong stump and rootball decay.

Completely remove all trees having stump diameters greater than about eight inches and backfill the cleaned rootball cavity with compacted backfill soil.

Zone 4

Cut all trees having stump diameters of six inches or less flush with the ground and treat the stump with a waterproof sealant to prolong stump and rootball decay.

Remove all trees having stump diameters greater than about six inches, install subdrain and/or filter systems, and backfill with properly compacted soil around the filter/drain system.

Zone 5

Cut all trees having stump diameters of about four inches and smaller flush with the ground and treat the stump to prolong stump and rootball decay.

Install a major embankment toe drain or subdrain system to lower the phreatic surface, filter, collect, and discharge embankment seepage. Incorporate major subdrain with tree rootball and stump removal where possible.

Remove all trees located beyond the toe of the downstream slope having stump diameters greater than about four inches. Install weighted filters and/drain systems in rootball cavities where seepage boiling and soil piping is likely to occur.

Tree and Woody Vegetation Growth Control Program

Many individual dam owners and small dam owner organizations are not financially capable of undertaking comprehensive dam remediation projects in one major construction contract. Therefore, they must undertake dam remediation programs in a sequential manner. The following sequential dam remediation program for controlling tree and woody vegetation growth provides the owner, regulator, and engineer with a reasonable opportunity to effectively evaluate the condition of an earthen dam and to prioritize dam remediation relative to observed dam safety issues.

1.	**First Year**:	Cut all tall grasses, weeds, underbrush, and trees and woody vegetation having stump diameters of four inches or less flush with the ground and treat all cut stumps with a waterproof preservative to prolong rootball and stump decay.
2.	**Second Year**:	Cut all trees in Zones 1 through 4 having stump diameters of six inches or less flush with the ground and treat the stumps to prolong stump and rootball decay. Keep all zones mowed and/or maintained to preclude renewed growth of previously cut woody vegetation. Repair most severe animal penetrations that exhibit seepage flows and/or cause unstable slope conditions on Zones 1, 4, and 5.
3.	**Third Year**:	Initiate comprehensive remedial dam repair investigations, analyses, and preliminary design. Remove all trees from Zones 1 through 3 having stump diameters less than about eight inches by cutting flush with the ground and treating the stump with a preservative to prolong stump and rootball decay.
4.	**Fourth Year**:	Finalize remedial dam repair design and begin construction of remedial repairs for all plant and animal penetrations that require special remedial dam repair design considerations.

5.   **Fifth Year**:   Finalize remedial dam repair construction and begin an
                                operation and maintenance program that will preclude the
                                need for future remedial dam repair associated with plant
                                and animal penetrations of earthen dams.

NOTE: **Earthen dams that exhibit severe dam safety deficiencies**

and dam safety issues that cannot be prolonged as a result

of potential imminent dam failure <u>are not</u> subject to the use

of this type of sequential dam remediation program!!!

Chapter 7
Economics of Proper Vegetation Maintenance

Regular maintenance on a dam, especially attention to trees and brush, is known to be critical to dam safety for several reasons (Tschantz, 2000):

- Overturning or uprooting trees causing large voids and reduced freeboard; and/or reduced cross-section for maintaining stability
- Decaying roots of dead trees causing potential seepage paths and piping problems
- Interfering with effective dam safety monitoring, inspection and maintenance for seepage, cracking, sinkholes, slumping, settlement, deflection, and other signs of stress
- Hindering desirable vegetative cover and causing embankment erosion
- Obstructing emergency spillway capacity
- Falling trees causing possible damage to spillways and outlet facilities
- Clogging embankment underdrain systems
- Cracking, uplifting or displacing concrete structures and other facilities
- Inducing local turbulence and scouring around trees in emergency spillways and during overtopping
- Providing cover for burrowing animals
- Loosening compacted soil
- Allowing roots to wedge into open joints and cracks in foundation rock along abutment groins and toe of embankment, thus increasing piping and leakage potential.

State and federal dam safety officials and other dam safety experts agree that trees have no place on dams. Federal agencies and some states do not allow trees to grow on dams. However, it is estimated that about a third of the nation's 77,000 inventoried dams have sufficient woody vegetation to hinder effective dam safety inspections (ASDSO, 2000; Tschantz, 2000). Most states require dam owners to remove trees and undesirable vegetation, but the cost of clearing and grubbing trees and restoring the dam embankment slopes and crest is often cost prohibitive for many dam owners, usually running into thousands of dollars. It would seem that regular control of woody vegetation and maintaining the surface on an earthen is relatively inexpensive, compared to removing trees on and repairing damage from neglected dams such as shown in Figure 1.

Figure 1.   Restored Downstream Slope on Fishing Creek Dam, Maryland (1991-92)

Likewise, it is important that owners maintain *desirable* vegetation on their dams on a regular schedule to avoid the expense of periodically removing undesirable heavy brush and mature trees. Early control is generally viewed to be the most cost-effective means of avoiding potential adverse effects on these structures from their continued growth (USBR, 1989). The bulk of maintaining a dam usually involves keeping the grass mowed and brush trimmed. An important question arises, ***"How much is a dam owner justified in spending to maintain a dam on a regular or annual basis to avoid having to bear the heavy cost of removing trees?"*** A correlative question then follows, ***"How often should a dam be mowed to control undesirable woody growth?"***

This chapter attempts to answer these questions, but there are many variables and site-specific factors which need to be considered. Some assumptions also need to be made.

Tree Removal Costs

The cost of clearing and grubbing a dam depends on the size and type of trees, growth density, total job size (i.e., number of acres of trees), location of growth (crest and/or both faces?), embankment face steepness, slope condition (such as degree of wetness or surface texture), degree and type of required surface treatment (backfilling, use of herbicides or bio-barriers, mulching, seeding, fertilizing, etc.), and regional labor and construction differences.

The reader is referred to Table 1 in Chapter 2 and Figure 2 below for unit area tree removal cost comparison experiences reported in a survey by eight state dam safety officials in different regions of the country. The survey data shows that the cost of clearing and grubbing trees and other woody vegetation varies widely within and among states, but generally ranges from about $1000 to $5000 per acre, depending on site-specific conditions (Tschantz, 2000).

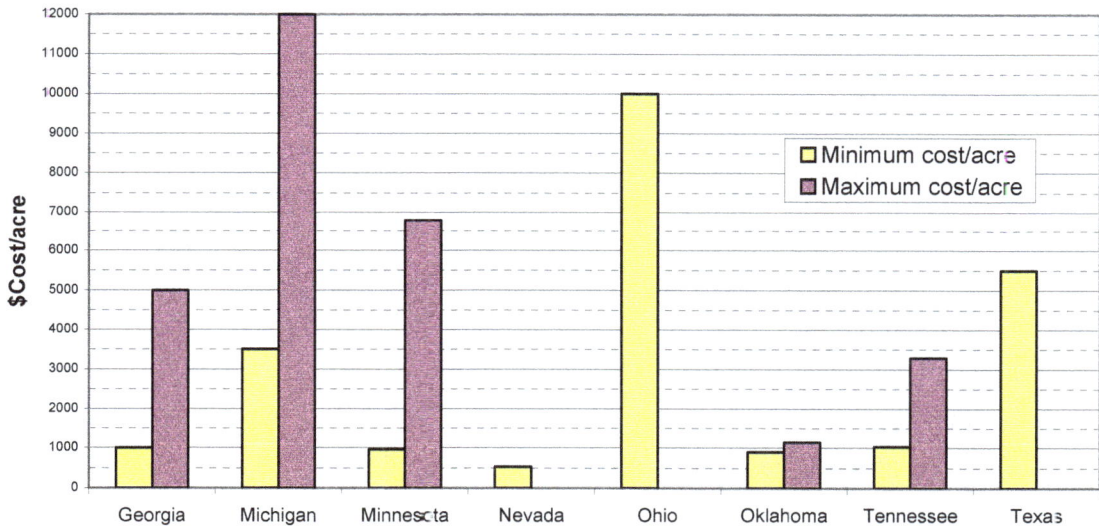

Figure 2. Comparison of tree removal costs per acre of dam surface area reported by 8 states.

These data compare favorably with the $1500 - $3000 bid price data for three Southeastern states discussed earlier in Chapter 2 for cutting trees, removing stumps and rootballs, and grubbing the area to remove roots for different dam conditions. While not included in the above Figure 2 chart data, Massachusetts' dam safety personnel reported in 2000 that, based on its own in-house experience, some local consultants and other sources, "broad area" tree removal costs ranged from $5000 and $6000 per acre or from about $800 to $1000 for individual 18-24 inch trees in their region. One dam safety official, from Tennessee, provided detailed cost data for clearing trees from seven dams in that state from 1995-1999. The cost for clearing and grubbing trees and for reseeding for one typical dam in 1998 is described for the reader in Table 1.

Dam Height	22.3 ft.
Length of Dam	830 ft.
Freeboard above Normal Pool	8 ft.
Density of trees ≤ 6 inches diameter primarily on downstream face	"Moderate"
Approximate surface area of downstream face	≅1.3 acres
Approximate dam face slopes	3H:1V
Amount of brush cutting	"Moderate"
Stumps grubbed out	Yes
Amount of hand work	"Considerable "
Total job cost for clearing, grubbing & reseeding	$4275
Unit area job cost	$3290/acre
Year job completed	1998

**Table 1. Tree clearing/grubbing and reseeding cost for a "typical" dam
located in Fayette Co., Tennessee (Bentley, 2000)**

For comparison purposes, general sitework cost information is available from various construction
cost books.   General cost data for cutting and clearing out individual trees and for clearing wooded
area is shown in Table 2 from one source (BNi, 2001). Indices are normally provided for factoring
in regional cost differences. Other cost book sources provide detailed material, labor and equipment
requirements for estimating site clearing costs (Means, 2001).

Clear small size wooded area:
- Light density
- Medium density
- Heavy density

$3,607/acre
$4,900/acre
$5,880/acre

Cut trees & clear out stumps:
- 9 to 12 inches diameter
- To 24 inches diameter
- 24 inches and up

$290 per tree
$370 per tree
$490 per tree

Table 2. General tree cutting and clearing construction cost data (Bni, 2001).

Similar general tree clearing and grubbing, chipping, seeding, mulching and fertilizing data for
estimating construction costs in various regions of the country are also available from other sources
(Means, 2001; AC&E, 2002). For example, 2001 Means cost data gives tree cutting, chipping,

clearing, and grubbing costs for trees 6-inches or less to be $2975/acre and stump removal to be $1425/acre for a total unit cost of $4400/acre. For trees up to 12 inches the cost is $6925/acre, and for trees up to 24 inches the cost is S15,250/acre (Means, 2001). If burning is allowed, the cut and chip costs can be significantly reduced. Hydro or air seeding, including seed & fertilizer is estimated to be 35¢/square yard (about $1700/acre) (Means, 2001). Mulching would add to this cost.

Maintenance Costs

For most dams, maintenance means keeping the crest and dam embankment slopes mowed and trimmed. The cost of mowing a dam depends on many factors, including geographical location, accessibility, condition of slopes as discussed above, degree of public use and desired aesthetics, type of vegetation and frequency of mowing. Cost also depends on whether the work is done directly by private owners, subcontracted commercially, or done by in-house state or federal maintenance crews. Table 3 summarizes these factors. The availability of slope mowers as illustrated in

Table 3. Factors Affecting Dam Maintenance Cost	
• Region of country	• Embankment slope steepness
• Type of ground cover & vegetation	• Mowing frequency
• Accessibility to dam	• Local labor costs
• Surface condition	• Type of maintenance provider
• Size of job (surface area)	• Degree of public use; aesthetics

Figure 3 illustrates the use of a slope mower for easing the burden of maintenance for state and federal agencies and for other multiple or large dam owners.

Figure 3. Example of slope mower (Terratrac© photo used with permission from AEBI North America, Inc.)

Most public works dams usually get mowed at least twice a year, in the early fall and late spring. Many subdivisions, homeowner associations, and/or residential developments typically mow dams, located in high-visibility areas, about once a month to every six weeks. One geotechnical consultant, who specializes in embankment dam rehabilitation, uses a "rule of thumb" mowing estimate of about $100 per acre with a minimum fee of $200 to $250 per mowing job (Marks, 2000). 1998 bid prices for mowing general right-of-way areas along East Tennessee highways averaged about $32 per acre, with a range of about $28 to $38 per acre for four jobs (TDOT-Region 1, 2000).

The U. S. Corps of Engineers, Nashville District, furnished recent annual mowing costs for three District dams, including some proximate recreation zones, having total mowing areas ranging from 8 to 27 acres. The average mowing cost for these three dams was about $55/acre and ranged from $43.42 to $78.24/acre (Corps, 2000).

The Tennessee Valley Authority furnished similar estimated in-house annual mowing cost data associated with general dam safety grounds maintenance activities for its dams. However, TVA's annual cost data included labor, supervision, slope mower fuel, parts, equipment, etc. and averaged slightly over $600/acre for 31 saddle and main embankment dams with a cost range from about $45 to $2000/acre (TVA, 2000).

A dam owner is advised that, in addition to mowing cost, the total annual maintenance expenditure should also include the expenses of dam inspection(s), minor repairs and rehabilitation of various structural components, removal of obstructions from emergency and service spillways, and other safety or operational costs associated with maintaining a dam.

Example maintenance cost analysis

The following example illustrates a rational procedure for answering the two earlier questions: 1) how much should a dam owner spend yearly to maintain a typical earthen dam to control trees and woody vegetation growth while avoiding bearing the cost of removing mature trees at a later date?

and 2) how often should an earthen dam be mowed to maintain acceptable ground covering vegetation? Maintenance expense in this example is for mowing only. Assumptions for this example are as follows:

1. **Dam Description:**
   - Length = 900 feet
   - Crest width = 15 feet
   - Embankment slopes (upstream and downstream) = 3:1 (horizontal to vertical)
   - Height = 35 feet
   - Normal pool = 10 feet below crest
   - Nearly vertical end abutments

2. **Economic Analysis Assumptions:**
   - 30-year project analysis period
   - Annual rates of return rates = 4, 6, 8, 10, and 15%
   - Zero annual inflation on recurring costs

3. **Maintenance Assumptions:**
   - Assume that 10-year old brush and trees are mature enough to significantly hinder effective inspection. Trees of this age can reach in size from 6 to 8 inches in diameter, depending upon species, tree density and other environmental conditions
   - Mowing costs = $100 per acre (with a minimum fee of $250 per mowing)
   - Trees can grow on all exposed upstream and downstream embankment slopes and the crest of the dam
   - Assume tree removal, including clearing and grubbing, costs = $2500/acre
   - Seeding & mulch not included in surface restoration costs.

Economic Analysis Calculations

Charts have been prepared and attached at the end of this chapter as a tool in helping to estimate mowing areas (or tree stand estimates) for different dam configurations. Chart 1 can be used to determine dam embankment slope area in acres for four slopes ranging from 1.5:1 to 3:1 (horizontal to vertical) and for dam lengths of 200 and 500 feet. Linear interpolations and ratio extrapolations can be made for other slope configurations and dam embankment lengths, respectively. Note that when determining the area of an upstream embankment slope, the equivalent dam height entered into the chart is the vertical distance between normal pool and crest elevation. Chart 2 is used to estimate dam crest area for three convenient lengths; crest areas for other actual dam crest lengths can be calculated from direct ratios. A self-guiding Chart 3 is provided to allow for small abutment area reduction corrections to be estimated and applied to slope area determined from Chart 1.

For the assumed example dam given above, make the following computations:

1. Use the attached charts to estimate total mowable and potential tree-covered dam area:
    (a) Downstream Embankment Slope (35 ft. high, 3:1 slope, 900 ft. length):
        A_1 = 1.28 acres x 900/500 = 2.3 acres (Use Chart 1; no abutment area reduction correction*)
    (b) Crest (15 ft. wide, 900 ft. length)
        A_2 = 0.17 x 900/500 = 0.31 acres (Use Chart 2)
    (c) Upstream Embankment Slope (10 ft. high exposure, 3:1 slope, 900 ft. length):
        A_3 = 0.37 x 900/500 = 0.67 acres (Use Chart 1)
    (d) Estimated total dam area to be restored ≈ 2.3 + 0.3 + 0.7 = 3.3 acres

2. Estimated 10-year cycle clearing and grubbing job costs, over a 30-year analysis period, starting with end of 10th year:

    Total Estimated Cost = 3.3 acres x $2500 per acre = $8250

    * For this example, the abutment slopes are assumed vertical or 0°, but total slope area reduction for a 30° abutment would be only ≈ 0.25 acres (see Chart 3).

3. Find the annual break-even cost balance between mowing and recurring clearing and grubbing, using the sinking fund factor (SFF), assuming 4, 6, 8, 10, and 15% discount rates for a 30-year period. A sinking fund is an equivalent annual amount to be set aside and left to grow at a certain interest rate into a specified amount at the end of a predetermined time period.

It is assumed in this example that mowing and clearing and grubbing costs do not change over the 30-year analysis period and that the dam safety inspections are not hindered for up to 10-year tree growth. By assuming a zero inflation rate for these costs, the results of this exercise are not dependent on the selected period of analysis; therefore, the annual values are valid for a 50- or 100-year period as well as for a 30-year period.

- Annualized clearing and grubbing cost = $8250 x (SFF, i, N years)

    where the SFF = $i/[(1 + i)^N - 1]$

        and i = discount rate (expressed as fraction)
              N = time period, in years, between tree clearing and grubbing

- Mowing job cost = 3.3 acres x $100/acre = $330

- Equivalent number of mowings per year = (Annualized clearing & grubbing costs)/($330 per mowing)

The following Table 4 shows that the annualized clearing and grubbing costs and equivalent number of annual mowings varies somewhat with the discount rate. For this example, at a 6% discount rate, this dam owner would be able to justify about two mowings per year at $330 per mowing to avoid having to shell out $8250 every 10 years for clearing the dam of trees and woody vegetation. The owner could afford to mow once or twice a year, even at a relatively high 10%

Assumed discount rate, i	Annualized 10-yr frequency clearing and grubbing cost	Equivalent number of mowings per year
4%	$687	2.1
6%	$626	1.9
8%	$569	1.7
10%	$518	1.6
15%	$406	1.3

Table 4. Annualized Cost Comparison for Assumed $2500 per acre for a 10-Year Cycle
Clearing and Grubbing Payout.

discount rate. By mowing on a regular basis, the owner would also realize side benefits of a more aesthetically pleasing dam -- one that would be viewed more as a community asset than a liability, be accessible and inspection friendly, and be less attractive to unwanted burrowing animals.

If $5000 per acre or $16,500 for 3.3 acres, rather than the $2500 per acre and $8,250 per job, had been assumed for tree clearing costs over the 10-year cycle control period, the justifiable annual costs for mowing would double for the same discount rates. For this higher restoration cost, the owner would be justified to mow 3 or 4 times per year, depending on the cost of money. The following Table 5 illustrates this assumption.

Assumed discount rate, i	Annualized 10-yr frequency clearing and grubbing cost	Equivalent number of mowings per year
4%	$1374	4.2
6%	$1252	3.8
8%	$1138	3.4
10%	$1036	3.1
15	$ 813	2.5

Table 5. Annualized Cost Comparison for Assumed $5000 per acre for a 10-Year Cycle
Clearing and Grubbing Payout.

For a more conservative 5-year tree growth cycle and a $2500 clearing and grubbing cost assumption, the annualized clearing and grubbing costs would be $1523, $1464, $1406, $1351, and $1224 for the same discount rates, respectively. The corresponding justifiable mowings would be 4.6, 4.4, 4.3, 4.1, and 3.7 per year. Similarly, justifiable mowings for an assumed $5000 clearing and grubbing cost would double the justifiable mowings to 9.2, 8.9, 8.6, 8.2, and 7.4 per year. Figure 4 compares 5 and 10-year annualized costs for $2500/acre clearing and grubbing payouts.

Equivalent cost in number of mowings/year

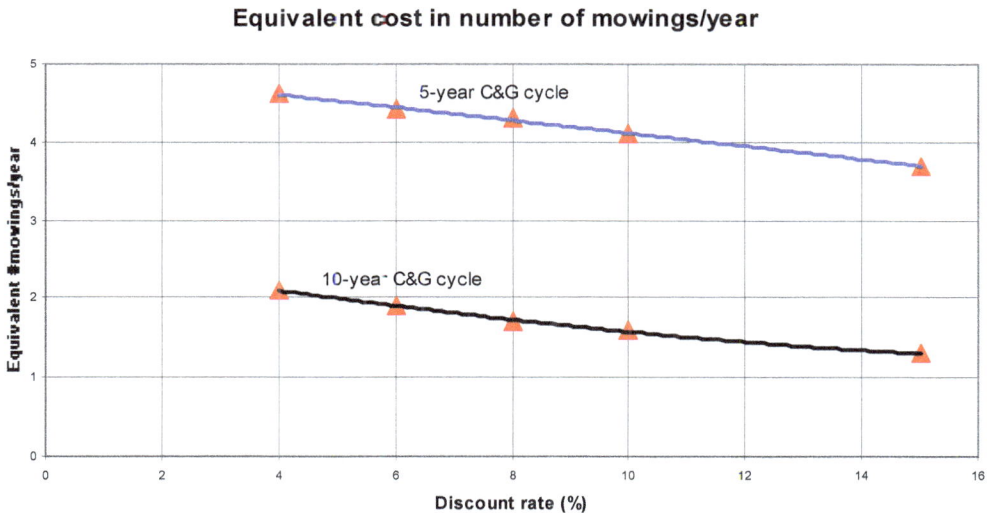

Figure 4. Comparison of Annual Tree Clearing and Grubbing Costs for 5 and 10-Year $2500/acre payouts.

Realistically, unit area costs would likely be reduced substantially for more frequent clearing and grubbing or bush hogging of smaller growth. Obviously, the above values will be different if the costs are assumed to escalate each year. For example, assuming a modest 3% annual inflation factor results in an increase in the clearing and grubbing cost from $8250 to $14,900 for a 30-year analysis period.

Summary

Cost data obtained from the private, state and federal sectors show that dam maintenance and tree removal and dam restoration costs can vary widely, depending on several factors.

It has been demonstrated, by way of example and reasonable cost assumptions, that dam owners can economically justify mowing their embankments 2 to 8 times a year, depending on local factors and costs, to prevent trees and other woody vegetation from maturing to a point that could compromise dam safety and require major capital outlays. It appears extremely economically efficient for dam owners to control woody growth on at least an annual basis, to avoid the large cost of removing mature brush and trees every 5 to 10 years and to comply with state inspection requirements.

So, how much should a dam owner spend on maintaining his dam? At least enough to keep it mowed and trimmed a couple times a year – probably something in the neighborhood of $500 to a $1000 annually for most dams, if contracted. Keeping a dam mowed a minimum of twice a year does not appear to be an unreasonable financial burden for most small dam owners. A dam owner must understand that spending a few dollars on annual vegetative maintenance and upkeep, such as mowing, will pay dividends over the long run for an asset (and potential liability) such as a dam.

References

1. Architects, Contractors, and Engineers (AC&E), Guide to Construction Costs, Division #2 – Sitework & Demolition, Cyber Classics, Inc., 2002.

2. Association of State Dam Safety Officials (ASDSO), State Survey: Animal and Vegetative Impacts on Dams, Part I - Vegetation on Dams (7 questions), September 1999

3. Association of State Dam Safety Officials (ASDSO), State Survey, Percentage of Trees on State-regulated Dams (2 questions), January 2000.

4. Bentley, L., Memorandum: Cost of Dam Clearing on Seven Tennessee Dams, February 2000.

5. BNi Building News, General Construction 2001 Costbook, Sections 02110.01-02110.50 (Sitework), 2001, p. 18.

6. Marks, B. D., S&ME Engineering, Inc., Arden, N. C., Faxed communication on recent contractor-bid clearing and grubbing costs, February 23, 2000.

7. Means, R.S., Building Construction Cost Data 2001, Site Preparation Section 02230, 2001, pp. 42-43.

8. Soil Conservation Service, U. S. Department of Agriculture, South Technical Service Center, Fort Worth, Engineering Technical Note 705, Operations and Maintenance Alternatives for Removing Trees from Dams, April 1981, 8 pp.

9. Tschantz, B. A., Overview of Issues and Policies Involving Woody Plant Penetrations of Earth-filled Dams, Presentation and Proceedings, ASDSO/FEMA Specialty Workshop on Plant and Animal Penetrations on Dams, Nov. 30 - Dec. 30, 1999, 8 pp.

10. Tschantz, B. A., Current Problems, Practices and Policies on Tree and Woody Plant Penetration of Dams, paper presented at ASDSO National Dam Safety Conference, Providence, R. I., September 2000.

11. Tennessee Department of Transportation (TDOT), Region 1 Supervisor, Mowing contract bid data – 1998, March 2000.

12. U. S. Bureau of Reclamation, U. S. Department of the Interior, <u>Guidelines for Removal of Trees and Other Vegetative Growth From Earth Dams, Dikes</u>, and Conveyance Features, Bulletin No. 150, Water Operation and Maintenance, December 1989, pp. 1-3.

13. U. S. Corps of Engineers, Nashville District, email information furnished by D. Williams, March 2000.

14. Tennessee Valley Authority, Knoxville, TVA Maintenance Data, Email attachment information furnished by J. Morse, April 2000 and August 2001.

Economics of Proper Vegetation Maintenance

Chart 1. Dam face area for different geometries

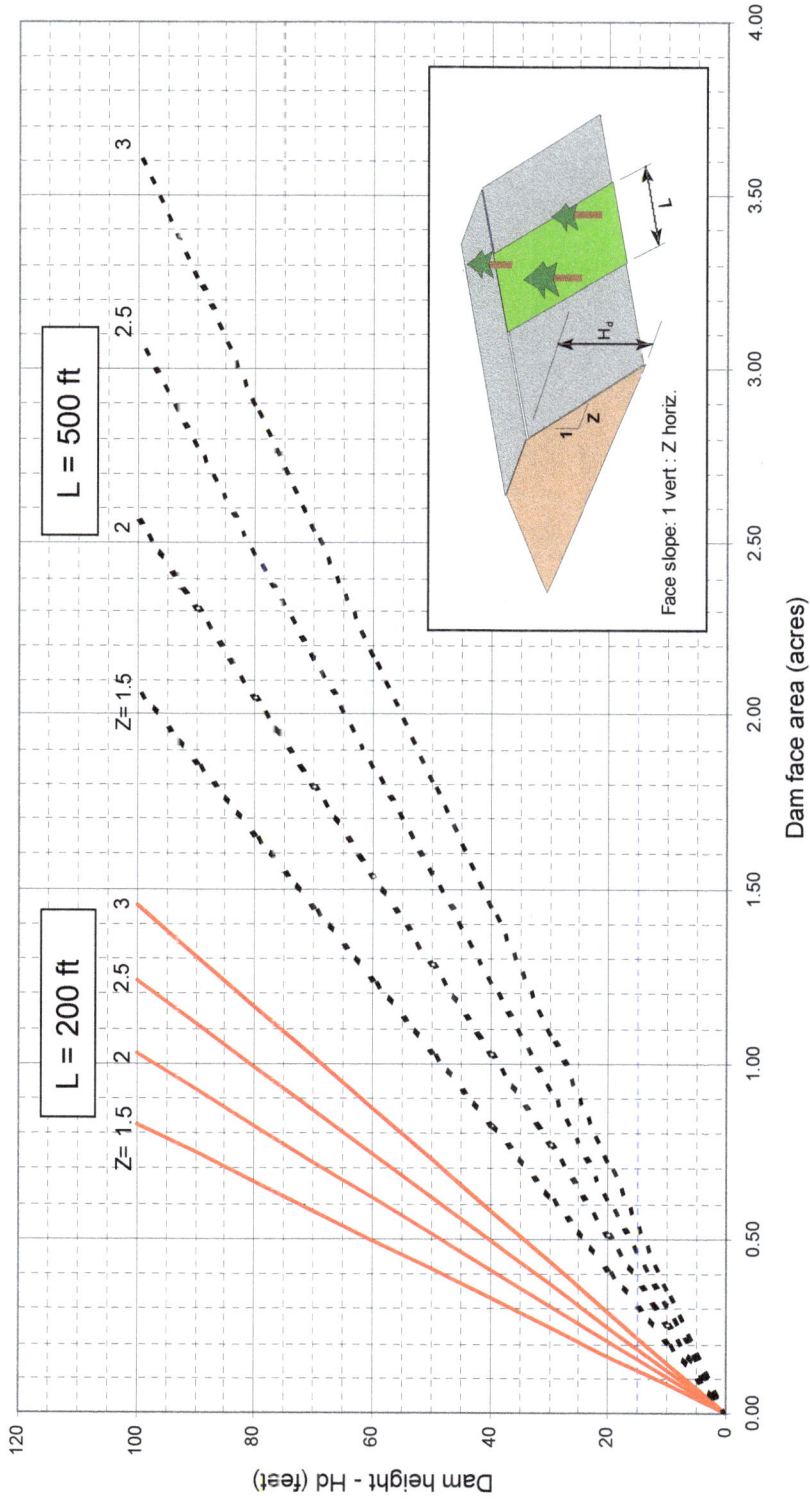

Face slope: 1 vert : Z horiz.

Economics of Proper Vegetation Maintenance

Chart 2. Dam crest area by width and length

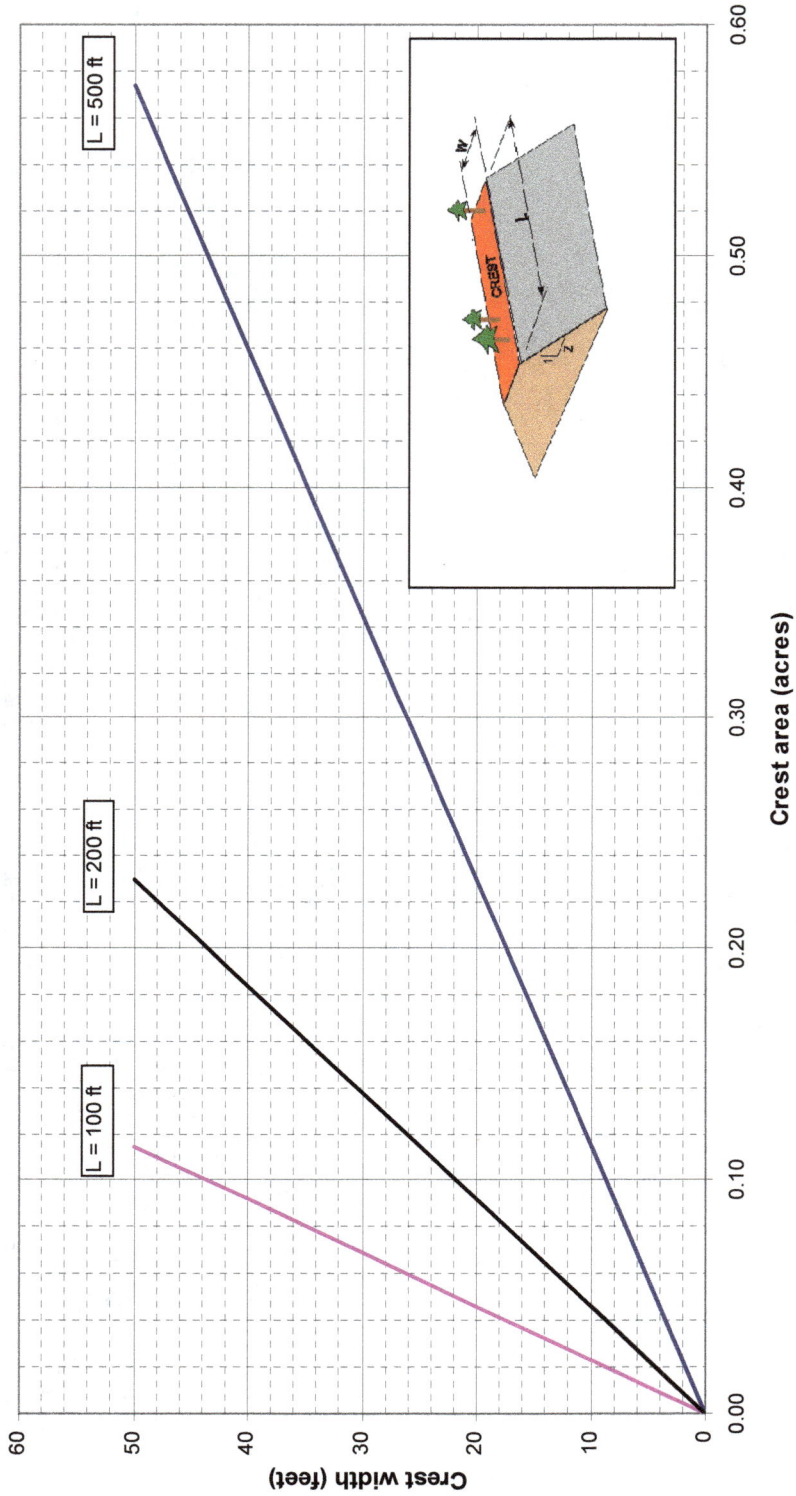

Crest area (acres)

Crest width (feet)

L = 500 ft

L = 200 ft

L = 100 ft

Economics of Proper Vegetation Maintenance

Chart 3.   Abutment area corrections vs. dam height for 2.5H:1V & 3H:1V side slopes, with 45° abutment angle and corresponding minimum dam crest length

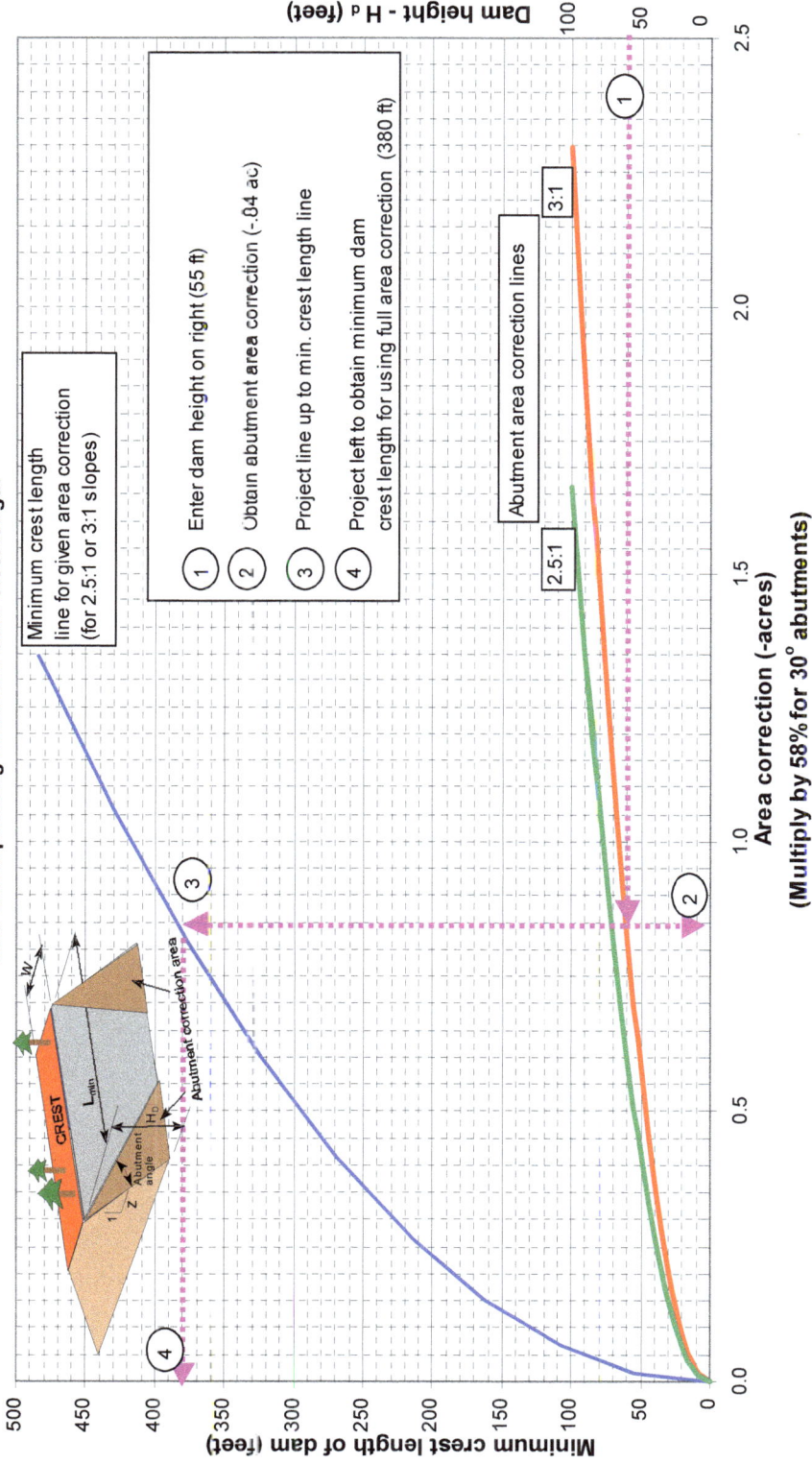

Dam height - H$_d$ (feet)

Minimum crest length line for given area correction (for 2.5:1 or 3:1 slopes)

Minimum crest length of dam (feet)

① Enter dam height on right (55 ft)

② Obtain abutment **area** correction (-.04 ac)

③ Project line up to min. crest length line

④ Project left to obtain minimum dam crest length for using full area correction (380 ft)

Abutment area correction lines

3:1

2.5:1

Area correction (-acres)
(Multiply by 58% for 30° abutments)

7-17

www.ingramcontent.com/pod-product-compliance
Lightning Source LLC
Chambersburg PA
CBHW051224200326
41519CB00025B/7234